Introduction

Much has been written about Harry Ferguson, his determination to improve efficiency in agriculture and the 'Grey Fergie', the iconic tractor that was the result of many years of hard work and development. Indeed, the first few pages of this book provide an insight into the great man's life and what drove him to create one of, if not *the* most iconic agricultural machines ever built.

It was, however, not the intention with this book to create the ultimate reference work for Ferguson tractors, but to highlight the sterling work done by enthusiasts aiming to preserve a true legend of British agriculture.

The Tractor England 20hp, or TE-20, has always been popular with collectors for many reasons. For some it was the first tractor they drove, for others it was the tractor that grandad had on the farm and, for a good few, it continues to provide a cheap introduction into tractor restoration.

It is, perhaps, inevitable that when the Standard Motor Company produced more than half a million TE-20s it would still be popular sixty years later – but for a tractor to continue to demand such a following is testament to the design and build quality of the little machine.

The versatility of the 'Grey Fergie' and its three-point linkage, coupled with the availability of petrol, tractor vaporising oil and diesel engines (among others) gave the sub-30hp Ferguson a fighting chance against rival manufacturer's larger machines – but it's almost certain that Harry Ferguson could never have envisaged the success his tractor would enjoy both at home and abroad.

Having achieved great things with his ground-breaking TE-20, it was obvious that Ferguson would look to build on the success by introducing a new tractor to the world. The FE-35 was an amalgamation of American and British technology with, in simple terms, Herman Klemm's dual-range transmission married to Harry Ricardo's 23C diesel engine.

The striking grey and gold livery was radical compared to the mundane 'battleship' grey of the TE-20, but the styling was still very much Ferguson. It was unfortunate that such a good-looking tractor could gain a bad reputation, largely through being a poor starter, but that is how the tractor is remembered.

Today, most can see past the FE-35's perceived faults and like it for what it is – a useful tractor that looks different to most others of the period. It's a tractor that many people loathe (thanks to its starting characteristics), but one owner that features in this book likes them so much that he owns three!

With enthusiastic clubs committed to preserving Ferguson's heritage, dedicated Ferguson classes at ploughing matches and a wealth of companies keen to offer a comprehensive range of parts for the tractors, it is unlikely that the popularity of the TE-20 and FE-35 will wane. They are relatively cheap to buy, simple to maintain and easy to transport – important factors to consider when buying a tractor. ■

GREATEST TRACTORS - **The Revolutionary Ferguson**

Published by

KELSEY PUBLISHING GROUP

Printed in England, United Kingdom by William Gibbons
of Willenhall, West Midlands
on behalf of
Kelsey Publishing Limited,
Cudham Tithe Barn,
Cudham, Kent TN16 3AG
Telephone: 01959 541444 Fax: 01959 541400
www.kelsey.co.uk

ISBN 978-1-907426-11-7

With thanks to:
Dave Beare, Gary Connolly, Barry Dinan, Andrew Hall, Kim Jackson, Peter Love, Bob McAllister, Dave Noel, John Pearson, Ben Phillips, Peter Phillips, Paul Smith, Alan Turner, Angela Walker, Bob Weir and Steve Wright.
Also a big thank you to the owners of all these fine tractors featured.

Contents

TELEPHONE
TELEPHONE
NO ENTRY
KENT COUNTY COUNCIL
ROADS DEPT
SKT 934
T 66

BLUEBELL RAILWAY
WAS STARTED IN 1959 BY A GROUP OF ENTHUSIASTS AND HAS GROWN OVER THE YEARS TO WHAT YOU SEE TODAY. THIS HAS BEEN
1882
NO PARKING
NO PARKING

Harry Ferguson at Abbotswood on October 26, 1950.

Harry Ferguson, genius

Alan Turner reviews his achievements and talks to some who knew him.

Henry George Ferguson was born on 4 November 1884, into a farming family at Growell, near Dromore in County Down, about 16 miles south of Belfast. Always known as Harry, he was one of a family of three girls and eight boys. Childhood was not an easy time for any of the children as father James was a member of the Plymouth Brethren and any books, other than the Bible, had to be sneaked into the house.

From an early age, Harry showed a flair for engineering and a general dislike for work on the smallholding and soon went off to join his brother Joe who ran a garage on the Shankill Road in Belfast and to attend night classes at the local technical college.

It was here he first met McGregor Greer and John Williams, who would come to have a great influence on his life. Harry's reputation for mechanical flair brought customers to the garage, but his restless mind was already moving on to embrace the new technology of aeronautics. He was soon heavily involved in building various monoplane aircraft, most of which eventually staggered into the air but all too frequently crashed back down again. Many of his flying experiments were conducted at Magilligan Strand, in Derry, the longest beach in Ireland.

Eventually, at Newcastle in 1909, Harry achieved a recognised success by claiming a £100 prize for being the first man in Ireland to complete a flight of three miles.

Safely back on earth, Harry moved on to start his own car business, May Street Motors, a venture that was partly financed by Greer and Williams. It was the start of an intense period in Ferguson's life, although it was not all work. He kept busy racing cars and motorcycles and the garage business gained a reputation for its principal's specialist knowledge of carburettor equipment.

There were other distractions, too. In spite of the agnosticism that was the reaction to his religious childhood he still became involved on the fringe of Irish politics. In 1912, apparently, he was helping the gun-runners of the Ulster Volunteer Force.

The following year he married Maureen Watson, who was to become an unfailing support to him for the rest of his life.

The World War inevitably brought changes and the garage business, now trading as Harry Ferguson Ltd, began selling imported tractors. However, the agricultural side of the business was far from successful due to the reluctance of farmers to accept any sort of mechanisation.

Around this time, Ferguson met another person who was to have a great influence on his life when Willie Sands joined his firm.

Doubtless the shortcomings of the tractors he was trying to sell did not escape his analytical mind but Ferguson first looked at the implements side of agriculture and by 1917 he had his own design of plough. Willie Sands was kept busy with engineering problems as he applied Ferguson's theories to the practical building of a prototype, while Harry transferred his attention to the ways in which implements were attached to the towing tractor.

Looking for a manufacturer to produce the plough soon brought him into contact with Ford through the American company's manufacturing plant in Cork. This led to contact with other American prospects and the result was a partnership with the Sherman brothers to produce ploughs at Evansville in the United States.

Absorbing though these other projects were, their pursuit did not mean that Ferguson forsook his other interests. The Ulster Grand Prix was a major car racing event that took place on a circuit comprising public roads near Belfast. It could only happen after Ferguson had helped guide the necessary legislation through Parliament, enabling the authorities to close the roads for the occasion. While no longer racing himself, when the annual event took place Ferguson's garage generally played host to one of the visiting race teams.

The agricultural side of the business received a bitter blow in 1928 when Ford responded to the economic upheaval in the United States by ceasing tractor production. Around the same time, Ferguson and his team had finally solved the problem of draft control that had been his personal goal for so many years. The solution, the three-point linkage, was to be embodied in a new concept – eventually, the Black Tractor – sketched out by Ferguson, McGregor Greer and Willie Sands in 1932. A prototype was built the following year.

Harry Ferguson and his aeroplane at Newcastle, Co Down, in July 1910.

The transmission specialist company, David Brown, was responsible for manufacturing the tractor but after a brief period, David Brown and Ferguson went their separate ways, with the David Brown concern eventually making its own distinctive line of tractors.

This left Ferguson clear to resume negotiations, this time with Henry Ford I, who was an agricultural enthusiast above his automotive interest. The two men had much in common and there were a number of meetings during the 1930s. Harry and his wife and daughter stayed at Dearborn as close friends of the Fords. The Ford company adopted Ferguson's principles for its own tractors and production went forward following the legendary handshake agreement. In 1941, the Sherman-Ferguson plough partnership finally came to an end.

While another World War shaped life, Ferguson stayed in the USA. Britain was exhausted in every way after the years of conflict and Harry returned to the UK in 1946 with a view to starting tractor production again. An introduction to Sir John Black of the Standard Motor Company paved the way and eventually Ferguson tractors were being produced at the Banner Lane factory in Coventry.

On the other side of the Atlantic there were big changes at Ford and, combined with the expiry of certain patents held by Harry Ferguson or his companies, Ford looked set to go its own way in both production and marketing of tractors, effectively leaving Ferguson high and dry. This brought out the tenacious fighting spirit that had seen Harry through all his trials so far. He issued a civil action against Ford claiming the colossal damages of $251,100,000.

The case began in January, 1948. Ford was in no hurry to settle, presumably feeling that it could grind Ferguson down financially. However, as the case dragged on the David and Goliath scenario raised public sympathy for Ferguson and provided him with the publicity that was always a spur in any of his endeavours. Eventually, in April, 1952, the two sides backed down and an agreement was reached.

By then Harry Ferguson was involved in another passion – Abbotswood, the country house he had bought near Stow-on-the-Wold, Gloucestershire. From here, he conducted his business and issued many of his famous long letters covering any topic on which he had an opinion. Inevitably, he also got involved with the fringes of politics. The Ford case and the fact that nearly 80 per cent of the tractors sold in the UK were Fergusons had given him gravitas as a businessman, but his political opinions were usually politely received before being dismissed, usually because of their naivety.

He became involved in Dixon-Rolt Developments, a company formed by racing drivers Tony Rolt and Freddie Dixon to develop a four-wheel drive concept vehicle called The Crab. This was initially intended as a race car, but Ferguson suggested the idea offered far more potential for road vehicles. He financed the company, but a ➥

The Irish Grand Prix on 6 June 1931 at Phoenix Park, Dublin. Harry Ferguson is on the pit wall wearing the arm band.

Harry Ferguson cleaning weeds out of the River Dickler, at Abbotswood, in 1948.

difference of opinion with Dixon soon had the company relaunched as Harry Ferguson Research in 1950.

In the same year Ferguson took on farmer's son Peter Warr as an employee on the Abbotswood estate, where he eventually rose to become estate manager. To date, working for the Ferguson family and its successors, is the only job that Peter has had. Harry may have been of pensionable age at the time, but his restless energy seemed undiminished to Peter, especially his enthusiasm for automotive matters.

Peter recalls his own involvement in many of the vehicle and implement demonstrations held at Abbotswood. The Duke of Edinburgh and Stirling Moss were just two among the many people that Peter drove around the Gloucestershire venue in the furtherance of a project.

Some were truly innovative, like the Mule. This was a joint development between Harry Ferguson and Rex McCandless, a Belfast engineer who was involved in motorcycles. The Mule was a four-wheel drive chassis, powered by a Norton motorcycle engine with a fibreglass-body capable of carrying four passengers over all sorts of terrain. Peter recalls the Mule as an extremely versatile machine and they had many adventures exploring its limits at Abbotswood. Nowadays we call such vehicles quads, but this project pre-dated the concept by many years.

The military authorities were disinclined to pursue the concept and so the prototype vehicle remains the only example ever built.

In 1952, the company of Massey-Harris-Ferguson was created, when Ferguson merged his interests with that of the Canadian agricultural engineers. The following year, Harry finally relinquished his interest in the business and sold out. He was still left with the research company and its automotive and agricultural projects, although the arrangement with Massey-Harris-Ferguson meant he could not be commercially involved with agriculture for five years.

Harry's grandaughter, Sally Fleming, can recall occasions at Abbotswood during her early years.

She said: "When I was small, I remember some important guests coming to lunch at Abbotswood. After lunch and a walk around the beautiful garden I remember my grandfather spending a lot of time huddled over a small tractor and plough model. He was, someone told me, explaining the principles of the three-point linkage system connecting the tractor to the plough. I listened in. He was still totally absorbed by the engineering issues asking: 'What would happen if we changed this?' The guests were fascinated, as was I.

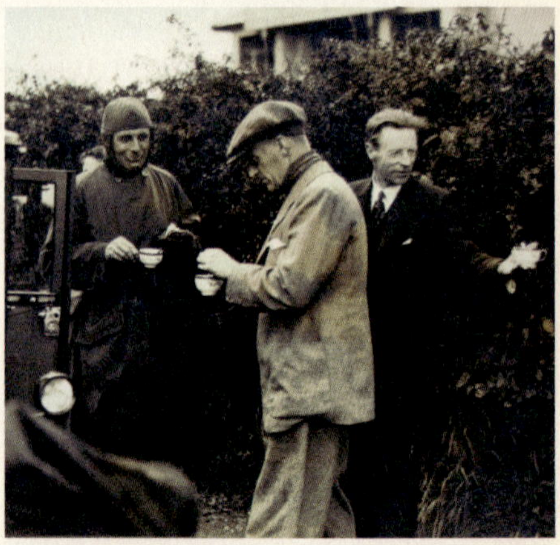
Lord Curzon, Captain Malcolm Campbell and Harry Ferguson having tea at Quarry Corner on 10 August 1928 before the TT race.

"I remember my grandfather being serious about anything he was interested in. He wanted the world to be a better place for ordinary people. He was single-minded, totally focussed and all his energies were devoted to realising his ideas."

Peter Warr also remembers Harry Ferguson with great affection. His employer's reputation for neatness and precision in every aspect of life was definitely not fanciful, including his minute-perfect punctuality.

Sally Fleming confirms: "He was punctilious about time-keeping. When he invited a guest to Abbotswood he would be ready and waiting to greet them at the front door exactly at the time for which they were asked. He did not like wasting time!

"Our family still regards being on-time to be very important. I feel horrible that I have let someone down, even when I am unavoidably late."

The butler organised the household at Abbotswood and had the enviable knack of being able to organise things to run like clockwork.

While Harry may have looked for precision in all things he certainly did not lose sight of the fact he was dealing with people. Peter recalls one of his lads missing an appointment with his boss by ten minutes. He had been playing football, but made the excuse that he did not have a watch. Harry Ferguson organised a trip into town for him to select and buy a watch and paid for it as well!

The staff members were on a twice-yearly bonus, which was awarded, or adjusted, for a general state of tidiness. Peter sums up Harry as a fair and considerate employer, who took a keen interest in his staff. He also appreciated initiative. Peter recalls an occasion when they attended a race meeting at Silverstone. Harry generously invited people to find his car in the Paddock, where they could enjoy a coffee. Unfortunately, the two flasks he brought were woefully inadequate against the number of invitations he issued. Peter managed to inveigle the catering concessions to keep supplying him with sufficient beverage. Later, impishly, Harry Ferguson acknowledged that he knew he could depend on Peter to find a way out of the dilemma.

Although perfectly formal for most of the time, when the two of them were together and no-one else was within earshot, he referred to Peter as 'Boy' and despite his many years away from Ireland he still retained a soft version of the accent.

In spite of the continuing advancement of his plans, there were still periods of depression for Ferguson. There can be few people in history who have been so innovative in so many fields of engineering. His enormous talents came at a price and he suffered from depression for much of his life, occasionally submitting to treatment for it.

On 25 October 1960, days before his seventy-sixth birthday, Harry Ferguson was found dead in his bath. A post-mortem

Harry Ferguson at Abbotswood in 1948.

revealed an intake of a large quantity of barbiturates. The man who gave agriculture its single greatest advance was no more. The jury at the Coroner's Court returned an open verdict.

Harry's death was an enormous shock. Peter can still remember the occasion vividly, when everyone at Abbotswood was stunned by the news and the staff wept openly. Peter is convinced the death was accidental, as he remembers Harry's enthusiasm at the time for the design of a new range of tractors.

The passing of Harry Ferguson was not the end of Peter's involvement with the family. He continued to work on the estate but then worked for Betty, Harry's daughter, after she married Tony Sheldon, and joined them when they bought King's Manor, a property on the Isle of Wight. This was taken over by their son, Jamie Sheldon, who continues ➥

Harry Ferguson's wonderful Costwold home at Abbotswood.

Harry Ferguson looks on as William Campbell removes a TE-20's brake shoes. Note the use of the Ferguson tractor jack.

to live at the house, while part of the farm has been converted to a small museum, a project that Peter, now in semi-retirement, has looked after for some years.

The museum can be viewed by appointment. There are examples of most of Harry Ferguson's landmark designs, including a model of the aeroplane as well as both earlier and later versions of the three-point linkage. From later years, there is The Crab, the four-wheel drive, four-wheel steered car that was the development basis of the Ferguson four-wheel drive system. This, of course, is still the principle of all the four-wheel drive vehicles sold nowadays.

Another major exhibit is the P99 Formula One car – the only four-wheel drive car of that type ever to win a race. The McCandless-Ferguson Mule is also in the museum. It is complete, but has not run for some time and the fibreglass body shows evidence of those enthusiastic demonstrations.

There are details, including engines, from the saloon cars that were also in development. Ferguson Research even pioneered the hatchback concept. Supporting all this is a huge stack of memorabilia, with many engineering drawings, photos, papers and letters.

The correspondence covers exchanges of views between Harry Ferguson and various statesmen and leading figures in industry in Britain and abroad. The work continues in trying to get the mass of material in some sort of order.

Sally Fleming notes: "There is a wonderful core of enthusiasts who are devoted to, and passionate about, Ferguson tractors. I have never been to an agricultural show where a Little Grey Fergie has not been a star attraction. The Ferguson Club and The Friends of Ferguson Heritage are working hard to perpetuate his memory."

With the engineering advances that were made in the twentieth century it is hard to imagine anyone in this century achieving anything like the advances made by Harry Ferguson.

Sally Fleming agrees: "I think he has, however, yet to receive a wider public recognition both for his contribution to farm machinery design and production, and for the innovative techniques he used to market his machinery. There is a place for Harry Ferguson in the history of twentieth century marketing and branding."

Asking Peter to turn his mind back to the time he spent working directly with Harry Ferguson, is there any overall impression? Without hesitation he sums it up: "It was a privilege to have worked in the presence of genius." ■

The ensemble, as used by Harry Ferguson.

A piece of history

Alan Turner tracks down the centrepiece of that Ferguson-Ford meeting.

Most Ferguson enthusiasts – and many of those who favour Fordson – will be aware of the often-reproduced publicity photograph showing Harry Ferguson and Henry Ford at a meeting on the Ford family estate at Dearborn, Michigan.

It is one of several taken at their famous pre-war 'handshake agreement' whereby Ford produced tractors to Ferguson's designs.

In most cases, the two are seated either side of a table and between them is a model of a tractor which could run on a short length of track. Apart from its usefulness in demonstrating Ferguson's products, it is also possible the model was made to support the application for patents on Ferguson's designs in the United States.

That model, or something similar, eventually became a more familiar sight as facsimiles were used by the salesmen pursuing potential buyers with Ferguson products. The models were capable of a miniature, but dramatic, demonstration of the advantages of the Ferguson three-point linkage over trailed ploughs.

The models issued to salesmen in the UK were made by Mills Models, a specialist model engineering company based in Sheffield. A number were also made in the USA for a similar purpose.

In common with some other owners, the Henry Ford Museum in Dearborn claims to have the very one in the photograph, but I think I can present sufficient evidence to convince even the most sceptical that the original is definitely on this side of the Atlantic.

I was invited to the home of the model's present owner. He prefers to remain anonymous but was perfectly happy to show me the model.

The entire kit is contained in a case made of plywood with the drop-down access secured by nothing more high-tech than a couple of nails. Nowadays, the case is a sombre matt black, but older photographs show it once had a gloss finish. It is topped with an old carrying handle that has definitely seen a lot of use.

The case contains the tractor, a trailed plough, a plough mounted to the tractor by a three-point linkage and a trailer. The ensemble is painted Ferguson grey, naturally, except for the red-painted trailed plough. There is also a small cardboard box. Removing these items allows access to the two pieces of track on which the model is designed to run. Dowels in corresponding holes align the two sections of track.

The tractor's chassis and tin work are folded from sheet steel, with soldered joints. Other parts, such as the wheels and tyres, were simply bought out. In this case, they were standard Meccano issue and still inscribed with the maker's name.

The tractor is powered by clockwork. Turning the tractor's starting handle tensions the spring. The drive gears are also standard Meccano items. The shafts for the clockwork mechanism and the axles appear to be cut to length from mild steel with flats filed in the appropriate places as locations for set (grub) screws.

The front tyres are showing signs of distress and no longer make an accurate fit on their wheels, although the model is still capable of being operated. The seat is neatly formed and drilled, but the drawbar features several alternative holes.

The trailed plough – the villain of the piece as far as Harry Ferguson was concerned – is painted red. To prove the demonstrator's point, the model tractor draws the plough ➥

Harry Ferguson (left) and Henry Ford in one of the famous photographs taken during the 'handshake agreement' at the Ford family's Dearborn estate. The model tractor can be clearly seen between the two men.

along the track but the passage of the share is blocked by a white-painted piece of wood, representing a rock or some other obstruction to smooth ploughing.

This sudden halt then causes the model tractor to rear up convincingly. The same exercise can be repeated to demonstrate the advantage of the tractor-mounted plough, located by the fabled three-point linkage, where the same obstruction will simply take weight off the tractor's back axle, eventually causing the rear wheels to spin.

The tractor-mounted plough is formed from steel; the share is welded or brazed and secured to the frame with some rather mangled round-head screws. The bottom links are formed from wire, the top link assembly from strips of steel, filed to profile.

Another feature of this model that further emphasises its unique status is a trailer, which appears to be a later addition to the original ensemble and hitches to the tractor by a pin that drops into the tractor's drawbar.

The single axle is a piece of steel rod bearing two 'Juneero' wheels, which are interesting in themselves. Almost forgotten now, Juneero was a model engineering system that never achieved the status of Meccano. Following similar principles, it featured a versatile special tool that enabled the builder to punch, shear and form parts to his or her own specification.

On this trailer, the axle assembly is located on the trailer bed by a fabricated hinge, which was soldered slightly skewed and has been subject to some remedial work to bring the axle into the correct axis.

The pivoted axle enabled Harry Ferguson to demonstrate his theories on towing. Pivoting the axle assembly changes the angle at which the trailer meets the tractor.

The trailer drawbar is adjustable 'fore-and-aft'. This enabled Ferguson to calculate the ultimate drawbar length to give the maximum transfer of trailer weight to the rear axle of the tractor.

The whole model bears evidence of a hard life. The block that represents the obstruction when ploughing has obviously been struck many times. Many of the pivoting components are quite worn; it looks as if some items are scarred from trips to the floor. Modifications or repairs have been made, as evidenced by numerous marks from the application of hand tools and the patterned damage caused by clamping with unprotected vice jaws. This was a working model in every sense of the description.

The cardboard box reveals further evidence. It contains a number of spare parts. such as springs for the clockwork mechanism. There is also a small 'cabinet-handle' screwdriver, a bent axle, a damaged spur gear and some worn emery cloth (read on for the significance of these items). An American-sourced matchbox stores smaller items including a Meccano spur gear and some spare set-screws.

The tractor runs along two grooves in the track and parts of these have been faced with coarse glass paper near the 'obstruction block'. This, presumably, is to enhance the dramatic effect of the tractor rearing.

The drive mechanism is one of the features that make this model unique. Manufacturers of the later 'production' models adapted their own mechanisms, which made a difference to the shape of the tractor. They also fitted their own wheels and tyres that were modelled as genuine tractor tyres.

One truly distinctive feature of this model is the bonnet. It is not quite right as it meets the front plate (representing the radiator ➥

Demonstrating how the Ferguson plough is mounted on the tractor.

grille) at different angles from one side to the other. In some of the photos in the Henry Ford Museum collection, this variation is clearly visible.

An envelope, fragile with age, was in the case and this contained some revealing details about the model and the relationship between the two industrial giants.

After the war, Ferguson got back to business needing to fulfil an enormous demand for its products. Changes at Ford led to a souring of the relationship between the companies. Eventually, it ended in an enormous lawsuit that was finally settled for several million dollars.

A few of the pages of this fascinating document, written in December 1948, provide ample evidence that the situation between the two concerns had reached an acrimonious stage.

It appears Ferguson accused Ford of interfering with his model, perhaps almost to the point of sabotage. Ford responded by claiming to have improved the performance of the tractor by 150 per cent! The matter was in the hands of the legal department and counsel painstakingly listed the remedial work done.

The work on the tractor was carried out in New York. As this consisted of replacing a broken drive spring, adjusting various set-screws and oiling the mechanism, it would not appear to have been too serious. Ferguson claimed parts of the tractor had been damaged and this included a bent axle.

With the main lawsuit between the companies yet to be settled, it provides an interesting twist to this battle of industrial giants. In this instance, they were feuding over a rather crudely made model that relied on Meccano parts, which were cheap and readily available.

For performance enthusiasts, Ford claimed the tractor travelled 11 feet on un-carpeted floor initially but following their ministrations, this increased this to 25 feet.

So what happened to the evidence? Copies of the paperwork include a list of damaged parts written on Ferguson headed notepaper. It is also possible to check all the items mentioned as they were all contained in a separate envelope, also preserved.

As noted above, all the contentious items listed in the original legal document remain with the model. The small screwdriver, or an identical one, appears in several of the photographs of the Ford - Ferguson meeting.

The spares and tool kit – re-housed in a later cardboard box after the original became too fragile.

The model collection has been in the present owner's family for many years as his father worked for Harry Ferguson Research in a number of capacities and it was Harry Ferguson who gave him the model.

As more people become interested in the development of mechanised agriculture the stature of the model has grown over the years. When the owner visited the Henry Ford Museum in Dearborn he was soon in conversation with an interested curator, who supplied copies of some of the relevant photographs.

When originally received by the family, the case handle still had a luggage label tied to it complete with Harry Ferguson's home address in Stow-on-the-Wold. This is now carefully preserved, as is everything that was originally received.

For the time being, the present owner has no plans to sell and wishes it to remain a family heirloom.

The photographs prove that the original model was considerably different to the samples used by the dealers and salesmen, which turn up at auction from time to time. They inevitably attract enthusiastic bidding and a present catalogue estimate is around £1,500-2000. ■

The clockwork mechanism, built by Ferguson – improved by Ford?

Shades of grey

Living close to the Ferguson homestead has inspired Sam and Colin Taylor to seek perfection.

Living close to where it all began for Harry Ferguson is a major reason why father and son, Sam and Colin Taylor have taken a keen interest in all things that bear his name. Their home in the small village of Drumaness in Co Down is only eight miles from the Ferguson homestead at Lake Growell, Dromore.

The pair started off with one Ferguson TE-D20, which was their sole tractor for 10 years. They have now expanded their interest to nine different Ferguson/Massey Ferguson models, most of them in original workable condition.

These include a petrol vineyard TE-K20, another TE-D20 with 'banana' loader and a 23C-powered TE-F20 (the engine probably adapted from a Commer van) On the red side, the pair own an MF 130, a three-cylinder 35 and a hard-to-find petrol-powered MF 135 – which has a serial number starting with D, which some authors believe stood for "demonstrator."

They have also built-up a small collection of Ferguson implements, including the earthmover, reversible plough, high-lift 'banana' loader, trailers, post-hole borers, buckrake, saw bench, winch and sack lifter.

But one of the pair's most recent projects was to restore a 1942 Ford-Ferguson 2NAN.

Why a Ford-Ferguson? Colin explains: "I'm sure every Ferguson lover dreams of a type A/Ferguson Brown but their £12,000-£15,000 price tag makes them a little less affordable to the average person, so the Ford-Ferguson is the next model in line.

"Dad's first driving experience was on a Ford-Ferguson so this was another reason to buy one."

One of the appealing factors when this particular Ford-Ferguson came up for sale was that the engine had been overhauled and there was only a minor snag to be rectified.

Things are never that simple are they? When they tried to get the tractor to run ➥

The Ford-Ferguson's look was completed with the addition of Firestone single-rib front tyres and 10-28 closed pattern rears. These were purchased from the Miller Tyre Co in the USA and flown to Heathrow, from where they travelled the rest of the journey by road.

The Taylors' three-cylinder MF 35 loader tractor came in very useful when moving the chassis around during the spraying process.

there was a major tight spot as the engine cranked over. That set the trend and the engine was then pulled to pieces. Colin takes up the story.

"It was out with the non-standard bearing cap and poorly-reground crank, in with the new cap and re-surfaced replacement shaft. The engine also consisted of two chrome liners and two cast ones, which was not really the done thing when restoring, so they also had to go.

"They were pulled out and a new set of pistons and cast liners fitted. A new oil pump repair kit, 20thou oversize shell bearings and three core plugs sorted the bottom half out.

"The valves, which had the Ford logo on them, were in very good nick and deemed fit for re-use. The valve guides, along with their springs and a handy set of adjustable tappets were sourced from Helms tractor parts, a very helpful company in the USA.

"The governor was also past it so it had to have an overhaul kit fitted.

"After having the cylinder head skimmed and tested, the engine was rebuilt and new clutch pressure and drive plates and pilot bearing, it was then refitted to the tractor.

"A re-cored radiator and new water pump were added and the engine was then fired up and, thankfully, all was healthy."

Readers familiar with this tractor will be aware that its maximum road speed is a lowly 8mph and this was an area the pair wanted to improve on. They sourced an American Sherman step-up transmission and fitted it along with a new input shaft gear with helical teeth.

Colin recalls: "Care had to be taken when drilling the shift lever hole in the main gearbox casing as you only get the one chance at it."

Other components of the main gearbox were replaced and re-shimmed as necessary. The fitting of new gaskets completed the gearbox and the test drive showed a very healthy 22mph on a handheld GPS.

After making sure all entry holes and seals were well protected the engine was wrapped up in a heavy rubber mat and the chassis was sent for shot-blasting in nearby Ballynahinch. Stanley McMaster took care of the blasting process, as he regularly does bits and pieces for Colin. He then applied two coats of zinc etch primer.

The front axle centre pin and bearing were replaced and stub axles re-bushed and reamed to size.

All of the rear linkage friction points were skimmed and re-bushed with new pins and the addition of new ball ends to the linkage arms. New oil seals were fitted to the pto and half-shafts along with new bearings.

The steering box was next and, after inspection, it had the shell bearings and rope-type seals replaced. The original gears were in excellent condition.

New brake linings, resurfaced drums and re-bushed brake pedals sorted the brakes. A new swivel bush was also made for the clutch pedal.

A new hydraulic pump control valve was fitted and, along with a light honing of the cylinder and new rings, the rear end was complete.

It was then time for the paintwork and every care was taken to preserve any original grey paint. The inside of the brake back plates, underside of the tool box and air cleaner were polished up and sent to be analysed.

Knowing that an exact shade of Ferguson/battleship grey never existed and there must be at least 40 shades of grey, especially when you see a line of Fergies, attention was focused on getting as close to this particular tractor's original paint colour as possible, allowing for fading over 65-odd years. After several attempts the pair had a shade of grey they were happy with (slightly darker and more of a blue grey than the majority of the TE-20 range).

The tin work was sent to a professional refinisher to be repaired and prepared. Meanwhile Colin was busy with an angle grinder buffing the high spots around the tractor chassis with an abrasive disc. Then he applied 10-12 coats of high-build primer (wet sanding after every two) to give the desired smooth chassis effect.

When the pair were happy with this process the tractor was given three coats of two-pack primer and, after hardening, was wet sanded. Then four coats of Standox (Lada 624 grey – 1972) were applied and left to harden.

According to USA sources and any original books and pictures available, the rear rims, fuel tank and wiring tube had a cadmium-plated white finish to avoid corrosion, although it seems many were painted over at some stage. After a few late nights repairing the rims this finish was copied

After a previous sub-standard engine rebuild, a new set of pistons and liners was required to bring things up-to-scratch.

The numerous layers of paint have really paid off – just look at that shine!

more simply by applying a white/silver metallic paint.

Several smaller parts arrived from the States, including radiator cap, original battery cover and six-volt lights with wing supports that had been in a box at a Ford dealers for 50-odd years. Colin recalls: "What a relief it was to see them light up when we attached them to a battery!"

New gauges, wiring loom, cut-out, starter switch, seat cover, coil and resistor and distributor parts all arrived and "it was time thankfully for re-assembly, when you actually feel you're getting somewhere," said Colin.

The tin work was back and the registration number, owner's name and address and tractor's unladen weight were hand-written on the bonnet, which was the norm for Northern Ireland tractors at that time.

The repaired starter motor and dynamo were fitted, along with new fan belt, thermostat and radiator hoses. A new exhaust was made up and fitted, along with the repaired heat shield.

It was then time to run the engine. It started first turn and was charging with great oil pressure, according to the gauges.

Sam and Colin have decided to run the engine on petrol for now as there was trouble with the old Holley vaporiser leaking fuel all the time. As parts for it are unobtainable at present it was seen as the best option, so all-original petrol model parts were sourced and refurbished.

The tractor was almost complete and there was a decision to make on the tyres. Miller Tyre Co in the USA could supply the original Firestone single rib fronts and were reproducing the 10-28 closed pattern rears so the pair decided that, having come this far, they would not skimp and the tyres were flown to Heathrow and taken by road the rest of the way.

So two years after they started the job was done and more pictures of the tractor can be viewed on Colin's comprehensive website: www.myfergie.co.uk

With the tractor now complete who knows what will be next in line? It's a safe bet that it starts with the word Ferguson! ■

The combination of grey paint, chrome details and matt black tyres looks fantastic.

Safety demonstration

The Ferguson system revolutionised mechanised farming in the 1940s and is still used by modern tractors today. Brent Thomas shows us the demonstration model that was used to convince farmers of the safety and control benefits of the new three-point linkage and Ferguson plough, versus a trailed plough.

The demonstrator model's main purpose was to show the safety benefits of the Ferguson system. This was very important – many people had died on tractors when trailed ploughs caused the tractor to rear up after hitting an obstacle. Some of the national newspapers were beginning to monitor the death toll, particularly in America where Ford was given a score all of its own.

Notwithstanding this key factor, the 'lecturers' that were sent around with these models to make the demonstrations were also briefed to push the benefits of farming with tractors rather than horses. Aside from pointing out that 20 per cent of a farmer's land must be used to support horses, lecturers were to stress the shortcomings of the horse while working, when compared to a tractor.

This can be summarised as follows: the horse must be fed an hour before it starts work, his working day is limited, as is his speed. He is liable to sickness, high maintenance and required more housing space than a tractor.

It may be worth mentioning that both Harry Ferguson and Henry Ford hated horses. They believed that, with around 20 per cent of productive farmland supporting the horse, it was contributing unnecessarily to hunger and starvation.

While the first attempts at mechanised farming tended to produce large, heavy and expensive machines that required more than one operator, smaller tractors with implements controlled from the tractor were considerably more practical, especially for smaller farms.

Having made this point, the demonstration of the model tractor, which would have been set up in front of the audience, would finally begin.

The demonstration

Having successfully eliminated the horse as a competitor, it was now the turn of the trailed plough. The first thing to be clearly demonstrated would be the weight of the front of the tractor with both the trailed plough and the Ferguson plough.

Here a member of the audience would be invited to lift the front end of the tractor with each plough fitted. The difference in

Brent Thomas shows off the model before demonstrating the benefits of the Ferguson system.

The demonstration model packs away neatly in a sturdy wooden box with both ploughs secured for transportation. Half of the 'runway' can be seen in front of the box – this is the lid. The other half of the runway slides under the box.

weight was clear, proving far greater stability with the Ferguson System.

Another very visual way of clearly demonstrating this point to a larger audience would be to attach a long elastic band to the starting handle and demonstrate the amount of stretch necessary to lift the front end with both of the ploughs.

After winding up the spring (which powered the model) through the starting handle, the tractor with the trailed plough could be seen to rear up physically after hitting the obstruction. The front end of the tractor can also be demonstrated, by hand or elastic band, to be considerably lighter. Carrying out the same procedure with the Ferguson plough and the three-point linkage – and, guess what, there is no reduction in the weight or stability of the front end of the tractor.

The forward thrust secret

The trailed plough pulls down on the rear end of the tractor through the hitch point and because of its final drive, the torque reaction will tend to exert pressure to lift the front off the ground, clearly reducing the load on the front end.

The Ferguson three-point hitch plough balances the load taken off the front end of the tractor through torque reaction by the forward thrust of the top link. In practice this means that in the event of hitting an obstacle, the rear wheels are more likely to spin. This is clearly a much safer outcome than having the front end of the tractor rear up.

By not having wheels on the plough, the fingertip depth control level will also allow easy adjustment for variable soil conditions, thus efficiently allowing for the best cultivation results. The science can sound confusing to many, but the model clearly proves the point! ➡

The models were built by Mills Models of Sheffield.

The starting handle is used to power the model. This is wound clockwise while lifting the rear wheels.

The starting handle tightens up a spring inside the model which, in turn, drives the tractor forward.

If the spring is tightened up enough, the front end will rear up when the trailed plough hits the obstruction. It is also very light to lift.

When fitting the Ferguson plough, the three-point linkage is quickly and easily demonstrated.

The wheels are placed in the tracks on the 'runway' with the plough heading towards the obstruction.

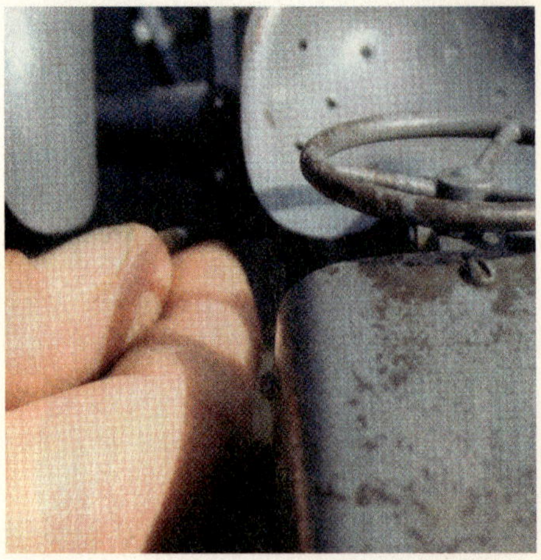

Above: The model has a hydraulic lever which can be adjusted to the height of the plough...

... enabling it to stay clear of the ground (below).

***Right:* When the Ferguson plough hits the obstruction, the tractor remains controlled and the front doesn't go light.**

***Below:* There were at least two types of demonstration model issued to Ferguson sales staff. The first of the two, and presumably the initial version, was basic in appearance and had no mudguards and little detail. It was rather rudimentary and wasn't as well proportioned as the other version. It is interesting to note that the ploughs, despite the obvious differences to the wheels, were remarkably similar and only the tractor's styling was altered between the two versions. The wooden carrycase also differs slightly, but it is unclear why these changes were made.**

Photos: Courtesy of Jim Russell.

The two tractors await restoration.

Two for restoration

Ben Phillips tells us about two early TE-20s he has acquired for a total price of just under £320. One Continental and one Standard-engined petrol/tvo model that are patiently waiting to be restored.

I can't resist buying a rusty old tractor! I see the potential of what it could look like after a restoration far sooner than I see the rust. This pair of rough-looking Ferguson TE-20s would look like scrap to the untrained eye; however, I can already see them in my mind looking like new.

It all started when my neighbour, Ken Briggs, rang me one lunchtime and said he had been to collect a Ferguson from a guy who lives next to a farm he owns in a local village. He was to take it to the monthly auction held at Stone, near Kidderminster. It had no reserve and no VAT and could possibly make a low price.

I rushed around immediately, hardly having time to put the receiver down. When I arrived I climbed aboard Ken's trailer to take a closer look and soon decided that if it went for reasonable money I would have it, as it clearly had the potential to make a nice tractor.

We roughly dated to 1951 as it looked to have a 12-volt electrical system – the smaller starter and the long bulge in the clutch housing easily identify this. The fuel tank suggested that it was a petrol/tvo model.

The tractor line-up at the auction: spot the difference!

The tin work was very good, although it had been painted green and there were a few dents and scratches in the bonnet. The 'T' piece would require a good hammering out to straighten things up. The grille was all present and correct but the rear wings were damaged and rotten in the usual places. I already had a spare second-hand set for such a model and, in any event, replacement wings are well made and not expensive to buy.

The tyres were more of a concern as the front ones were split and totally flat, while the rears were odd patterns with one looking like the original – but at least they both seemed to hold air well.

Auction day

I decided I would go and bid to see if I could get it. Unfortunately, on the day, work commitments prevented me from going so I asked Ken to bid for me. I set a price limit of £150 and was later told that I had bought it for £120. I was delighted! On finishing my jobs for the day I arranged to pick it up. Ken and I arrived at the sale ground to collect the tractor later in the afternoon and there she was, maintaining as much dignity as possible in her green and yellow colour, parked next to a big 1990s' John Deere.

Ken connected a rigid tow bar between the Fergie's top link and the towing point under the front bumper of his four-wheel drive truck. It was then a relatively easy job to push the tractor onto the trailer, strap it down and get on our way home.

Getting it started

The first challenge was to get the tractor running and that was my plan over the following weeks. I removed the plugs and gave them a good clean up. Then I sorted out the wiring and made sure fuel would run through the tap and into the carburettor.

I got a well-charged battery, connected it, turned the ignition switch on and engaged the starter via the gear lever... the engine turned over beautifully. I quickly reached for the choke and as I did, the Standard engine spluttered slowly into life, but stopped just as quickly. ➡

It's the green one with yellow wheels – they managed to come away with the right tractor!

These look like they could be the original tyres, and they still hold air.

The extra coating of paint may have helped, but the shell-type mudguards show typical wear and damage.

You can see here the 12V starter motor used on the petrol/tvo model. This differs in form to those used on earlier models.

Despite the chrome ring suffering corrosion, the gauge seems to be in good condition.

For about an hour I tried but could not get it to keep running. I pulled the carburettor apart and cleaned out the jets but this made little difference. Then, as I stood there scratching my head, Ken turned up on a completely unrelated matter.

Being the excellent engineer that he is, the carburettor was soon off and apart again. Ken identified the problem – the pin at the top of the float had seized and was not letting enough fuel through.

Once the pin was free the little TE-20 started and ran okay. I have a bit of a miss-fire to sort out, but nothing that will prove to be a major problem.

Double trouble!

Some time later I found myself trawling through the tractors for sale on eBay. I wasn't in the market for a vintage tractor, as I have enough work waiting to be done, but a particular tractor did catch my eye – an early Continental-engined TE-20. To top it off, the tractor was not too far away.

As it happened, I was more interested in making the most of the good weather and forgot about bidding for the tractor until after the auction closed. At the time I remember being a little disappointed, but two days later the tractor was re-listed – it had obviously not sold the fist time. I made up my mind that this time I would try and get it. I studied the pictures carefully and spoke to the seller to ask some questions. He mentioned that he could deliver it for a little extra money.

When it came to the final bidding, I seemed to be the only one interested and as a result I got it for just under £200. We arranged delivery and within days I had another tractor lined-up for restoration.

I believe from the casting marks that this model was built in March 1947. It has the small independent brake pedals and tiny wheel nuts that were phased out by 1948/9. The bonnet is not too bad, with a few bits of rot, but nothing problematic.

I noticed there were two odd-looking brackets bolted to the front axle and a tap was fitted to the oil return. I worked out this must have been for a loader and Ken confirmed that it would be a Horndraulic model, a popular implement in its day.

Everything on the tractor was seized so I decided to remove the cylinder head and fill the bores up with a mixture of diesel and old engine oil. This was easier than it could have

This 1947 Continental-engined TE-20 represents more of a rare find and an interesting challenge for a restoration, although it is still possible to source spare parts.

been – all the bolts that needed loosening only required gentle persuasion. The pistons were all stuck, but the bores were not displaying any wear at all.

I made up a diesel and oil mixture and poured it in. It disappeared quickly from three of them so it must be penetrating somewhere! Every now and then I add more of the mixture, which should free the pistons.

One day both of these tractors will be treated to the restorations they deserve and I look forward to having the time to really get started!

Buying tractors on eBay

I would advise anyone considering buying a tractor on eBay to go and have a look at it first and only pay when you pick-up the tractor, or have it delivered.

This is, of course, all the more important if you are buying an expensive tractor! ■

Front wheels have smaller nuts.

A 6V dynamo is fitted to the Continental-engined TE-20.

The engine with the rocker cover removed.

I now have the tractor home after the engine diagnosis. Due to a change of circumstances I re-assembled the engine in order to transport it. The front of the bonnet is undented but has some corrosion along the left-hand side, possibly where it was laying in a hedge for some time? I will attempt to repair it to retain originality.

New project, new problems

Andrew Hall discovers what work needs to be done to his Continental-engined Ferguson.

Those of us familiar with collecting tractors know that there are very few collectors who stop at one machine. I am no exception to this rule and rarely turn my back on a fresh restoration opportunity, having recently acquired a 1947 Ferguson TE-20 with Continental engine.

For anybody unfamiliar with the Continental-equipped models, they were the first of the TE-20 models manufactured at the famous Banner Lane works of the Standard Motor Company in Coventry. Despite the name 'Continental' the engines were imported from Detroit, as the more commonly-used Standard-designed and built engine had not been completely developed at the start of production.

My machine was offered to me by a friend and requires plenty of work to get it up and running again. At some point in the distant past somebody has painted the tractor in blue with black bonnet and wheel centres. Perhaps the tractor was used by a local authority or a private company and those were their colours.

The registration number is EEW 650 and, after a little investigation, I learned that it originated in the Huntingdon area.

Curious to know more of its history and perhaps how it received its black/blue colour scheme I contacted the County Record Office in Huntingdon and for a small fee they supplied a copy of the original registration entry along with adjacent vehicles registered at the same time.

The first owner was Hubert Milner of Yaxley, near Peterborough, and the registration date was 24 March 1947.

I have since made contact with a member of the family and the tractor was well remembered, having been used with a slightly later Ferguson TE-D20 working on celery crops on the black peat-like fen soil. The black/blue colours must have come later, after the tractor was sold. If anybody recognises the tractor and has any further information on this machine I am interested.

When acquiring a fresh machine for the collection it is all too easy to turn one's attention to that and to neglect existing projects and as I have my David Brown 2D restoration underway, this is where I should be concentrating my efforts! However I was curious to establish what was required to get the engine running.

Evidence showed that the tractor has not been run for many years and the engine was 'stuck' through lack of use. First task was to attempt to free the engine but, despite soaking the cylinder bores with diesel and using much effort to turn the crankshaft, it wouldn't budge.

I removed the cylinder head and discovered the diesel had drained down past the pistons. On inspection I discovered some small cracks in the cylinder block between the cylinder liners but, undeterred, proceeded to strip the engine totally.

After seeing no specific problems with the cylinders I thought the problem lay in the crankshaft and bearings region, but when stripped, again no positive reason for the seizure!

At this stage a complete clean of the cylinder block was undertaken in a professional wash tank as the engine sump contained a lot of brown sludge which would have reduced the oil flow around the block. In the final stages of cleaning I noticed a black line across the centre main bearing web of the block. This turned out to be a crack right through the centre of the block rendering it unserviceable.

So what now I wondered. Continental engines are not too common. Calls to various breakers showed no promise until I called Patrick Smith of Fergieland. Patrick specialises in all things Ferguson-related and has an extensive knowledge of the machines. He was able to supply a serviceable block at reasonable cost to me. It has now arrived and is as good as he said.

In the meantime I have temporarily re-assembled the engine and put the tractor back together as I couldn't keep it in the workshop where I began the strip-down. I will now have to fit the rebuild in with my other projects. In anticipation of the job I have been purchasing parts when I have seen them available. My intention is to get the tractor running and to a driveable state and use it for light work such as paddock topping and the occasional ploughing match, if possible. Cosmetic work will have to come later. ■

The tractor in as-found condition. A replacement seat came with the tractor and has been fitted. The black object placed on the gearbox is the starter motor, which was removed to test.

The tractor came with a nine-hole drawbar and stays, together with stabiliser brackets under each axle. The oil seal for the pto will need replacing. Note the 1⅜in pto adaptor.

To the casual eye the Continental engine is similar to the later Standard engine, but a point to note is the forward position of the oil filler and the horizontal ignition coil mounted above the starter motor.

The underslung exhaust has rusted away and fallen off, so is another part for the shopping list. The original Continental engine, such as this one, was designed to run on petrol and did not have a vaporiser fitted. Note the tie-bar running along the sump as the Continental sump is pressed steel, rather than cast iron.

Life in the fast lane

Bob Weir meets a racing legend's relative and his fine Fergie.

The Borders region of Scotland is a vibrant farming community, but also has a motor racing pedigree second to none. I drove to Jim Clark country to meet the great man's cousin, Doug Niven, and his recently-restored Little Grey Fergie.

With the exception of the legendary Stirling Moss, Clark is arguably Britain's most respected racing driver. Apart from winning two Formula One World Championships, he was also the first Briton to win the prestigious Indianapolis 500 in the same year.

Clark was a member of the Berwick and District Motor Club and began entering competitions at the tender age of 20. He eventually came to prominence as a member of the Border Reivers Racing Team (BRRT), led by two local enthusiasts Alex Calder and Jock McBain.

The team was based at what is now Doug's garage at the village of Chirnside close to the A1 and, although the original building has since been demolished, Doug still has lots of memories and memorabilia.

He said: "Having such a famous relative I suppose it was only natural that I would want to try and follow in his footsteps so I started motor racing in 1969."

Because he had a busy farm to run, Doug decided to stay closer to home and in 1970 chose to enter the Saloon Car Championships. Just like his legendary cousin he made an immediate impact and won the Scottish Saloon Car Championship at the first time of asking.

This was quickly followed by more success and, during the 1970s, he became the first driver to win the Scottish Championship in two successive years. After winning more than 100 first-class races throughout the United Kingdom, he eventually called it a day.

Being first for Doug had become something of a habit. Once he had returned to full-time farming, he started to introduce some ground-breaking innovations. The family's farm at Whitsome Hill near Duns, became the centre of a flourishing estate management business that is still going strong today.

He said: "I was concerned that farming was becoming less profitable, and wanted to take advantage of economies of scale. I decided to set up the Merse Agricultural Management Company Ltd (MAMCO) in 1985." The new company's first client was the estate owned by former Prime Minister Sir Alec Douglas Home.

When Doug had taken over the farm in 1964, Whitsome Hill covered 620 acres but as his new venture mushroomed in size, he was soon managing an impressive 7,000 acres.

Once he had consolidated this expansion, he turned his attention to newly-liberated Eastern Europe. Following the dismantling

Doug Niven aboard his Ferguson TE-A20.

During the 1970s Doug was one of the most successful sports car racing drivers in the UK. Photo courtesy of Doug Niven.

Doug celebrates with the checqured flag. Photo courtesy of Doug Niven.

of the Berlin Wall in 1989, Doug was one of the first British farmers to expand into Czechoslovakia. This meant commuting alternate weekends for the next five years to manage a 1,600-acre farm producing potatoes and cereals.

He said: "With the benefit of hindsight the proper infrastructure wasn't in place during the early years. Farming in Czechoslovakia required a lot of investment for a comparatively small return, so in the end we decided to concentrate our efforts full-time in the UK."

As if he hadn't got enough on his plate, Doug also found time to join LEAF (Linking environment and farming).

"LEAF was set up in 1991, and we became the first demonstration farm in Scotland," says Doug. "This was also a pioneering exercise as it was the first time a group of farmers, environmentalists, food suppliers and consumers had got round a table together to try and do something positive for the farming industry."

Back at Whitsome Hill Doug had come to an arrangement with Co-operative Farms, which took over the contract management agreement from MAMCO.

Not the sort of man to let the grass grow under his feet, Doug was one of the first farmers to pioneer the use of Global Positioning Satellite technology for yield and soil nutrient mapping.

He discovered that the 7,000 acres now run by the Co-op varies significantly from light sands by the banks of the River Tweed, to a heavy loam further inland. He has also set up another company called Precision Field Services, offering similar advice to local farmers.

In between estate factoring and consultancy work, Doug is vice-chairman of the RSABI (Royal Scottish Agricultural Benevolent Institution). He also sits on the council of the Scottish Agricultural College. He has even become something of a local celebrity and regularly broadcasts a farming programme on Radio Borders.

He said: "This started off just as a trial for a limited period as the new station lacked anyone with local farming knowledge."

That was back in 1990 and Doug has been doing it ever since. In addition he also finds time to write a column for the weekly paper *The Scottish Farmer* under the pseudonym 'The Gleaner'.

With all this activity going on at home and abroad, Doug admits he really hadn't paid much attention to restoring old tractors. But he hadn't reckoned on the magical lure of owning a Little Grey Fergie and when the opportunity to buy one came out of the blue he couldn't resist the temptation.

He said: "Like a lot of farmers I have always had a soft spot for the TE-20, and bought it for something of a hobby. From the age of eight, every school holiday I would drive the Little Grey Fergie whenever possible, doing jobs like harrowing and rolling fields that would not be allowed today."

Doug's 1948 TE-A20 (serial no: 27297) was bought in 1989 from a sale in nearby Kelso. He has never been too concerned about the tractor's history, but it is believed to have spent most of its life working on local farms. The attached plough is the original one used for many years at Whitsome Hill.

Despite his good intentions, the pressures of work intervened and the tractor was left to languish outside in all weathers for the next 15 years. Then in 2004, Doug finally decided to have it restored.

"I discovered that one of our farm workers Kenneth Allan was a keen tractor enthusiast and when I saw the superb job he had carried out on his own collection of Massey Fergusons I asked if he fancied having a go at renovating the TE-A20," Doug explained.

Fortunately for the tractor Kenneth decided to take up the challenge, despite the fact the Little Grey Fergie had already been christened 'Old Rusty' by Doug's grandchildren.

To begin with, Doug shied away from a full restoration and was just happy to see the tractor up and running again. But, as is often the case with makeovers, the job quickly gathered its own momentum and eventually turned into a full bottom-up restoration.

Doug, who is also a member of Friends of Ferguson Heritage, said: "The bodywork was certainly in a dire state as over time a lot of it had rusted away. This meant Kenneth had to rub the whole thing down before the tractor was sand-blasted and then repainted. At the same time I also agreed to fit a new set of tyres and mudguards." ➡

Despite water ingress in previous years, the engine burst into life after an oil change.

The TE-A20 is by no means concours, but a worthwhile restoration all the same.

Kenneth and Doug were particularly concerned about the condition of the petrol engine. Over time water had managed to seep down the exhaust and, when they tried to drain the engine, they found a lot of muck before they finally came to the oil. Yet much to their amazement, after an oil change and a bit of coaxing the engine spluttered into life. A testament if one was ever needed to the Little Grey Fergie's endurance.

It was always Doug's intention to try and preserve the tractor in its original state and, despite all the hardships the engine has endured, it was decided to let sleeping dogs lie. The unit is still in pretty much the same condition as the day it left the factory. But engine notwithstanding, there were plenty of other problems in store.

"I remember that Kenneth had a lot of problems sorting out the electrics," says Doug, who also owns a 1964 Ford Super Dexta and an original Ferguson trailer, which was used on the farm for many years and has recently been restored. The control box and ignition switch had to be renewed with parts supplied by Southern Counties Tractor Spares Ltd down in Chichester.

Kenneth then found that the engine was overheating, which meant having to buy a new radiator. At the same time the thermostat and cracked housing were also replaced.

In the end the whole job took 18 months, mainly because Kenneth works full-time on the farm and could only afford to spend one hour each morning working on the tractor.

But, like the best of stories, this one has a happy ending. Now that it has finally been restored Doug's Border Reiver has come in from the cold and is safely stored away in a nice warm garage. ■

The original Ferguson plough was used at Whitsome farm for many years.

The tractor awaiting restoration.

Restoring on a budget

There aren't many opportunities to buy a TE-A20 for £75 and so when Peter Phillips had the opportunity to do just that, he jumped at the chance.

My interest in tractors began in the 1950s when, as a teenager, I started to spend most of my spare time on a local dairy farm. The highlight of the weekend was to be allowed to drive the David Brown petrol/tvo Cropmaster up and down the rows of kale. After an apprenticeship on this job I was soon allowed to drive on other easy jobs and this soon led to being able to use the newer Ferguson TE-20. With the onset of the busy summertime I was soon involved in haymaking and hay carting – a time of many happy memories and summer days that we don't seem to have these days.

Growing up meant that spare time was increasingly occupied with other interests and messing about on the farm gradually declined. A job in accountancy meant that my only contact with tractors came through entries in farming client accounts, although at least I was able to widen my knowledge of farm machinery from this source.

Moving on some 40 years, my interest in tractors has been rekindled by frequent visits to some of the many country shows that take place in the Worcestershire, Herefordshire and Shropshire area.

The opportunity to go and look at a long-discarded tractor owned by one of my clients led very quickly to me recruiting my neighbouring farmer, himself an old tractor enthusiast, to take me and my son Ben to collect our newly-purchased (for £75) Ferguson TE-20.

Once the tractor was safely removed from the edge of a local forest, where it had lain for several years, to the safety of my back garden we were able to take stock of what we had. After referring to a Ferguson book we established that our machine was a 1949 TE-A20 petrol-engined model and that despite a seized engine, being covered in green forest slime and very rusty, it seemed to be all original and all there.

Ben's daily task was to go round all the nuts and bolts with a tin of WD-40 in the hope that they would undo if and when we needed them to. I looked closely at each part and besides the engine, the main problems appeared to be rust, a radiator full of oak leaves and wiring part-eaten by mice. It was clear that the only sensible way of restoring this machine to a decent condition was to take it all to pieces and clean, de-rust and repair/repaint everything.

We started by removing all of the tin work and the lift arms, followed by the starter motor, dynamo, radiator, fuel tank, carburettor and water pump – in fact anything that could be removed to leave the engine more accessible. Now the cylinder head could be removed to investigate what was seized.

Removing the rocker cover exposed signs of quite severe rusting around only one of the valves, a hopeful sign that getting the engine running could be easier than first thought. Removal of the cylinder head was an easy task, involving taking off the head nuts and carefully prising up the heavy head using varying thicknesses of timber pegs. This was done firstly not to damage the head and also not to damage the head gasket – the intention being to try, as far as possible, to use as many original parts as possible for the sake of both originality and cost. The head was lifted a little at each corner until it was removed, to reveal that the only ➡

The strip-down is in full-swing with Ben looking ever the professional restorer!

serious rust was around number three cylinder. The bore was lightly rusted and the pushrod stuck solid, so we filled the cylinder bores with WD-40 and left it all to soak.

Having reached this stage I decided that to be in a position to obtain any information and advice in the event of further problems, I should join both the NVT&EC and, as a Ferguson owner, the Ferguson Club. By coincidence, the latter club was at that time based only 16 miles from my home. Both organisations issued interesting periodic magazines and provided contact with other tractor owners.

The engine is stripped down and painting begins, with Ben concentrating hard!

As things turned out, contact with other Ferguson owners made my restoration much easier and more enjoyable as it was so easy to solve problems by being able to quiz someone who had first-hand experience of today's problem. The workshop manual can appear somewhat daunting to a non-technical person like myself; in fact one young owner of a beautifully-restored TE-A20 actually suggested discarding the manual and working by instinct – perhaps easy for him as a tractor mechanic!

After topping-up the cylinders daily for several weeks, to try and un-seize the engine, I was recommended to gently tap the top of each piston with a 2" piece of timber and a lump hammer. This treatment soon had the desired effect and one momentous day, after standing idle for so many years, the engine could be turned over using the starting handle – the pushrods were free and we were making some progress at last. To be able to inspect and clean up the rusty bore(s) it was necessary to remove the sump and the pistons – not very difficult on the tvo and petrol-engined TE-20 but apparently much more difficult on the diesels.

With the pistons out we were pleased to find that the bores were in pretty good shape, needing only a delicate rub-down with fine emery paper. The pistons only needed a set of bottom rings, the big end bearings were passed as fit for service and the engine was soon put back together and ready to run.

In between the engine jobs all the smaller components had been taken to pieces, thoroughly cleaned, painted in primer and received a top coat in the correct shade of Ferguson grey. By this stage we had a good idea of the 'new' parts we needed and started to track them down.

It soon became clear that in spite of being so old, almost any spare part is still available, finding out where being the hardest part of the operation. A specialist firm, the Old Twenty Parts Company from Derby, were able to supply the correct spark plugs, distributor cap, points, radiator cap, fan belt and a reprinted owner's instruction book and workshop manual.

As a total rewire was necessary, instead of buying a new loom I purchased the various pieces of cable in as near to the original colours as possible. I rewired the tractor in accordance with the manual – an operation that probably took longer than fitting a new loom, but one that taught me a little about motor electrics and saved a few pounds too!

As the dynamo, coil and control box were in a rusted state and needed specialist repair I decided that replacement was the sensible option. These parts appeared rather expensive from the vintage tractor suppliers so I brought them from a local electrical repair shop. Painted grey it is difficult to tell that they are not the originals. While dealing with this part of the restoration I also changed the original 6-volt electrics to a 12-volt system, mainly to save the cost of an expensive 6-volt battery when I had more versatile 12-volt items available and to ensure easier starting as the Ferguson lives outside at my exposed property.

As the tin work was all in remarkable condition and only needed rubbing down, rust-proofing, undercoating and a top coat this had been done in-between other jobs and on wet days when outside work on the tractor was not possible.

Gradually all the parts were refitted and the machine started to take shape again, now in as near to new condition as was possible to achieve with my limited facilities and skills. Even the brush-painted tin work looked reasonable.

While investigating how to go about dealing with the paintwork, it was apparent that traces of the original registration number existed on both the rear axle centre housing and below the grille bars. As we did not have a log book, this evidence of the number was to prove vital in enabling us to retain the original number and so avoid having a 'Q' plate.

Not the best environment for restoring a tractor, but it worked for us!

After gently cleaning the faded letters it was possible to take photos to connect the number with our tractor. By reference to the serial plate beneath the steering wheel we knew the year and approximate month of manufacture. An enquiry to the local County Council Archives (in the area where the number was first issued) produced a copy of the page for April 1949, which included our tractor. ➡

Nearly done and looking resplendent, despite being hand-painted.

The finished article. Not bad, considering Peter is an accountant!

This information, the photos we had taken and the details we had found out about the history of the tractor and its previous owners (incomplete) were sufficient to convince the NVT&EC to recommend that the registration claimed by us was genuine and that a new log book should be issued by the DVLA at Swansea.

Very soon only a few jobs were left to be completed and this was when problems arose. One of the last tasks was to fit the new dynamo; seemingly a simple operation except that there was no way the 12-volt dynamo would fit with the 6-volt bracket. After some research I discovered that a different type of dynamo would be needed – a part that proved impossible to obtain from any of my 'new' Ferguson friends. One of my clients, a machinery dealer, gave me the number of an associate of his in South Wales. A phone call soon located the vital part and my client arranged to collect it for me and then leave it in the dog kennel at home in case nobody was about. Some days later he turned up at my house with the dynamo, complaining that I hadn't told him the kennel was still occupied by the dog – an Alsatian!

Within days I was ready to see if, after almost ten years, the engine would start. After a few attempts I called in my neighbour, who is something of a mechanical expert and had given me a lot of advice during my 'apprenticeship'. After some deft adjustment of the timing he soon had the engine running smoothly and everything seemed complete. An appearance in the vintage tractor section at the small show in the next village at last seemed possible. When we started it was our aim to appear at this show, which was then five months away.

One rear tyre had been losing air for the entire five months, so my neighbour offered to take it away and look for a puncture. We took the wheel off and were just wheeling the tyre away when the tractor rolled off its axle stand and sank into the soft ground. Getting the tractor safely back up took a valuable hour but at least nobody was hurt and this had been the only accident.

After a frantic evening wire brushing the inside of the wheel rim, the morning of the show dawned sunny and warm and after a quick run around with the touch-up brush I set off for the next village. This was the first time I had driven a Ferguson for many years and the tractor's first journey under its own power for ten years. We had the only TE-20 at the show and it attracted a lot of attention, not least from many of my farming clients who wondered how an accountant could possibly successfully restore an old tractor!

With the renovation completed it was time to take stock and without doubt it was one of the most enjoyable exercises I had undertaken. I met a lot of interesting people, learned a lot about both engines and tractors (especially Fergusons), found a perfect relaxation away from my daily accountancy work and, at the same time, gave my teenage son an interest to occupy his spare time. I recommend tractor renovation as a fascinating and absorbing hobby. ■

What a wonderful place to work!

A life in Provence

Dave Beare offers something interesting from a vineyard.

Of all the Ferguson tractor models seen in this country the rarest are arguably the TE-K, TE-L and TE-M20 Standard-built narrow-track vineyard tractors. They're easily picked out in a line-up because they look strange alongside normal Fergies.

Narrow Fergusons were introduced in 1946 with a track width of 42 inches instead of the standard 48 inches. The first TE-B Continental petrol-engined model was intended for horticultural use by fruit or hop growers and the like when a normal Ferguson was too wide.

Major alterations to a narrow tractor included shortened rear axle housings, half-shafts, a different front axle and modified radius arms. The rear linkage was also changed, lower links were cranked, check-chains shortened and a different levelling-box used.

The 1,966cc 23.9hp Continental engine was abandoned in Britain in 1948 once the Standard Motor Company's 1,850cc four-cylinder petrol unit was ready for production. It originally had 80mm x 92mm bore and stroke but was superseded in 1951 by a larger 85mm bore, 2,088cc unit which gained an extra 5hp as a result.

Ferguson TE-C20, TE-E20 and TE-J20 narrow tractors were fuelled by petrol, tvo and lamp oil respectively; diesel-engined versions were only made in France by Standard-Hotchkiss. British models continued to be made in small numbers until 1956 when narrow tractors were abandoned.

One other narrow Ferguson model was built between 1948 and 1952 by Scottish Ferguson agent Reekie Engineering of Arbroath, which first modified a Standard Ferguson by reducing the track down to 32 inches for local raspberry growers. It proved rather popular, with up to 500 being produced before Coventry decided there was money to be made and produced its own variant. Reekies usually have a special bonnet badge and often a blue stripe painted round the bonnet.

Even the 42-inch narrow tractors were too big for some uses such as wine growing, raspberry and sugar cane cultivation. Ferguson was aware of the success of Reekie's ultra-narrow tractors and introduced its own vineyard versions, the TE-K20, TE-L20 and TE-M20 in April 1952.

Changes were even more radical than for the 42-inch tractors; the track setting was 37 inches, giving an overall width of just 46 inches, and wheels were smaller in diameter – 24-inch rears and 15-inch fronts, thus also lowering the overall height of a vineyard Ferguson.

Otherwise, modifications were similar to a narrow TE-C20 – shorter axle housings, half-shafts and cranked linkages, with an additional casting fitted at the front of the engine to move the radiator forward. ➥

Press photo of an early Ferguson TE-C20 narrow petrol version. Note hubcaps, which were fitted to avoid damage to crops by wheelnuts.

Narrowing the front axle reduced the excellent turning circle of a standard-width Ferguson, so moving the radiator forward gave better clearance for the front wheels on full lock. The bonnet had to be lengthened by four inches.

France would be an important market for Ferguson tractors but initially strict customs and import regulations, put in place after the end of World War 2 to encourage local manufacturers, made imported ready-to-work tractors prohibitively expensive to sell.

Harry Ferguson did the time-honoured thing and put together a deal with French car and military vehicle manufacturer Hotchkiss to establish the Société Standard-Hotchkiss at St Denis outside Paris where, from 1953, French Ferguson tractors were assembled.

Initially, major components such as engines were imported but the new company was soon able to make most of the tractor from French-sourced parts: engines were made by Hotchkiss-Delahaye, electrical systems came from Ducellier and so on.

French Ferguson tractors were offered in similar ranges to British ones – normal was the standard 48 inch agricultural model, étroit was the narrow model and vigneron was the vineyard version.

France had no tradition of using tvo, having a more enlightened tax structure for fuels than that imposed on Britons by the Treasury, so two power units were offered, the Standard petrol engine and a French-built diesel.

This created a new model series, the TE-F20 (as the UK), TE-G20 (narrow) and TE-N20, the diesel vineyard tractor. This one, naturally, became very popular with vignerons all over France.

In 1956 these models were superseded by the FF-30 series: Hotchkiss-Delahaye had coaxed 30hp out of the engines but little was changed except the type numbers to FF-30GS (standard), FF-30GE (narrow) and FF-30GV (vineyard).

By now these tractors were totally made in France but they only lasted until 1958 when the faithful old Ferguson was dropped in favour of the Massey-Ferguson 835, built at a brand-new factory in Beauvais. Some 60,000 Ferguson tractors were assembled or made at the St Denis factory.

One vigneron Ferguson has escaped the usual fate of many tractors in France, that of being dumped under a hedge when replaced by something more modern. It has been saved, appropriately enough, by a Briton, Ian McLatchie.

He said: "This little Fergie came into my life some 10 years ago when I bought an old mas (local name for a building or farm) in Provence.

"Out in the country up a 1km-long unmade track, surrounded by vines, my building was in need of renovation (it had electricity, but no water or drains) as it once housed the horses and mules that were used in the vineyards, but now was used to house the little Fergie, their replacement!

"When I first saw the house with the estate agent I knew I wanted it, but I wanted the tractor as well. The agent was adamant that this was not possible as he had been told that the elderly owner had had the Ferguson from new and it was still being used in the vineyard, which was not being sold with the mas.

"The purchase was a lengthy process, mainly caused by the difficulty of finding water, which was conditional to the purchase. Three bore holes were drilled to 100 metres but no water, so my only choice was to have mains water trenched in over my neighbour's land, through his vines. Patrice would only allow this once the leaves had dropped and they'd had their winter prune, some 10 months' wait. This gave me many opportunities to bring up the subject of my wish to buy his tractor but the answer was always no.

"Almost a year later the water had been connected and the purchase of the house was completed at the lawyer's office. All parties have to sign so the vendor was present and, while saying our goodbyes, Patrice (the vendor) brought up the subject of the tractor. He said that because he lived some 5km away and had no garage, would it be possible for me to garage it? In return I could use it if I was careful. I was sad that I couldn't own it but very glad to have it available to use, so I happily agreed to it being left in my custody.

"Sooner than I expected I was astride the tractor because, when the first delivery of building blocks arrived, I received a call from the driver saying he was trying to turn round at the start of the track to return to his yard, as he had seen enough to know that his lorry would not fit up the track.

"I asked if he had a smaller vehicle to make the delivery, but he didn't. I told him to stay where he was so I could come down

Factory photo of the TE-L20 narrow vineyard petrol/tvo tractor introduced in April 1952.

Interesting comparison between a standard agricultural Ferguson and a vineyard model.

Ian McLatchie drives through the vineyard where the Ferguson vigneron spent its working life.

with the tractor and a trailer to load the blocks on to. The little Fergie loyally served me transporting all the building materials in this manner, in addition to carrying out its normal duties in the vineyard using a cultivator, spraying and towing trailers for the harvest piled high with grapes (a 10km round trip to the co-operative), as it had done all of its life.

"Embracing my new life and happy to have a break from working on the house, I would often help Patrice out in the vines but at first I was rarely allowed to drive the Ferguson as it was still his baby. However, over the years our roles reversed and he was more than happy for me to carry out some of these tasks with my wife Claire joining in, secateurs in hand, for the laborious task of grape-picking at harvest time.

"One day three years ago Patrice came to my house to tell me he'd decided it was time for him to retire while he was still in good health and enjoying his life. He was nearly 70 and had started work on the land when he was 14 but, sadly, neither his son nor daughter was interested in taking on the vines, so would we like to? It was not something that we had ever considered so we asked for time to think it over.

"There were so many sleepless nights and hours of discussion with Claire, but we finally decided that although we'd enjoyed the occasional work with Patrice, the idea of total responsibility for the vineyard was too much to commit to. It was a very difficult decision. Now we watch a new owner working the land with a much newer MF 354 and, sadly, we have no involvement.

"Patrice very kindly gave me the little Fergie on his retirement, but it had to be relegated to living in the open as we had turned its garage into our dining room!" ■

A Reekie-converted narrow Ferguson with blue band and extra bonnet badge. Note curved radius arm on front axle.

Un petit Ferguson

John Pearson introduces his French-built Ferguson TE-K20.

Introduced in 1952, Ferguson's vineyard models are, at first glance, just a narrow version of the standard tractor but when you take a closer look, some major differences are revealed.

To allow an extra narrow (37in wheel track) and low profile, smaller diameter wheels were fitted (24in rear and 15in front). As well as modifications to the axle housings, half-shafts, radius arms, steering arm and linkage, an extra four inches in length was achieved which, in turn, lengthened the bonnet. The concave mudguards are also distinctive.

The French-built Ferguson tractors, though displaying the Ferguson badge on the bonnet, were described as Massey-Harris-Ferguson on the identification plate and had their own serial numbering system.

My TE-K20 featured here is number 37030, dating from about 1953, but this number will also be found on a standard TE-20 of 1948!

***Above:** This original Ferguson disc plough on the TE-K20 tractor was used in the French vineyards before planting the vines. The plough was specifically designed for use on stony or hard ground of the type found on many vineyard slopes.*

***Above left:** Although it was sold in France when new as Ferguson, the TE-K20 is described on the serial plate as Massey-Harris-Ferguson.*

***Oppsite page:** The TE-K20 vigneron, as well as being very narrow, is lower and longer than the other Ferguson models and utilises the 85mm bore Standard Motor Company petrol engine.* ■

Technical specification

Manufacturer:	Standard-Hotchkiss, St Denis, France
Engine:	Standard Motor Company four-cylinder, overhead valve, four-stroke
Power:	28.4hp
Fuel:	Petrol
Displacement:	2,088cc
Electrical system:	12 volt
Cooling:	Water
Transmission:	4 forward, 1 reverse
Linkage:	Category I
Maximum speed:	19.5km/h
Lift capacity:	3,748lbs (1,700kg)
Weight:	1,240kg

The black Fergie

Peter Love takes a look at a black Ferguson tractor that is very different to many of its counterparts.

I came across Gerry Marmion's black 1952 Ferguson TE-D20 four-wheel drive for the first time some years ago at the hugely-popular first Ferguson gathering, right by Lake House, Growell, Dromore, County Down.

That day you couldn't get near the black four-wheel drive creation. I marvelled at the intelligent workmanship that had gone into the make-up of the tractor and I am sure Harry would have approved of it, if he was still alive. As like anything Harry did, Gerry's tractor had a special 'spirit' to it that certainly makes it stand out.

Number of conversions

I have been lucky enough to have seen the black four-wheel drive tractor at a number of events north and south of the border. At perhaps the biggest rally of its kind now, Fingal, which takes place at Rathmooney, Lusk, County Dublin, I was able to take a set of pictures of it and talk to the owner in more detail and take notes.

At the same time Gerry was being besieged by a raft of Ferguson enthusiasts who wanted to know more.

One must not forget that a number of Ferguson tractors were converted in working times using Selene axle kits for use in the hilly terrain of Devon and Wales, for example.

However, technology had moved on when Gerry undertook his 4x4 project. His son Brian, who made an excellent job of keeping up with Harry Ferguson's intention of having the tractor in 'shiny' gloss black, and Gerry's brother, Brian, helped him. As they are readily available, the use of a Land Rover Series III front axle was a must as a starting point, which in turn, harnessed Gerry's design ability that became part of the fun of building the tractor.

The Series III axle housing and shafts had to be cut down by 175mm. The axle mounting brackets were then MIG welded on, extending up to the former axle beam area, with suitable modifications as required. This also applies to the steering pivot arms

Gerry Marmion's Ferguson TE-D20 4x4 Mk2 looks the part at the Fingal rally.

Above and above left:* *Gerry has done a great job getting everything to work together in harmony. It's not easy marrying Ferguson and Land Rover parts together, so credit where credit is due!

Below:* *The transfer box looks neat, tidy and well engineered.

and track rod arms that are well structured to the front hubs.

Mysterious transfer box

The prop-shaft then comes off the front differential and heads neatly towards the transfer box that is all incorporated in the fabricated sandwich box placed between the gearbox and the rear transmission.

The pto engagement lever had to be shifted across to the other side, which wasn't such an easy job as you might think.

Essentially, the 'mysterious' transfer sandwich housing features an extended main shaft with a sprocket on the end. This then connects via a sturdy drive chain to a lower drive sprocket that is connected to the prop-shaft via bearings and a flange.

The art in all of this is the front axle sprocket ratio, remembering that the standard front 19-inch wheels were discarded and 16-inch secondhand Benford dumper wheels were used.

These are fitted with traction grip 8-ply 7.50 x 16 tyres and need to run approximately 10 per cent faster than the rears, which are on the standard 28-inch rims.

Gerry must have got it all right as he reports no wind-up whatsoever. However, free-wheeling hubs have been fitted on the front, so when on road work the four-wheel drive can be disengaged, thus saving on tyre wear and fuel economy.

Not for the purist

The tractor now features front wings, which are ideal for stopping all the mud and water spray from landing on the engine and sundry items. Suitable modifications have been incorporated on the footplate with checker plate that extends back to the rear axle housing and special, more practical, steps are also to be seen.

The wings are pure MF 35 and feature some stylish, fabricated brackets and lights fitted virtually parallel with the top of the wings. There is no standard Ferguson pan seat here – Gerry turned to the opposition by fitting a luxurious Ford 6X Series seat covered in black, not blue.

More power required

So what is it like to drive? Well, when I asked the question, I was told the tractor could do with more power under the bonnet particularly on a long drag as the extra weight certainly tells. Perhaps a TE-F20 diesel version would have been better?

I have, however, driven John Crossle's Roadless Dexta 4x4 in Northern Ireland some years ago on the National and that lacked power at times, even though it was a diesel.

Perhaps the tractor could be called a deluxe or custom model of the TE-D20, but whatever you might think of it, the engineering that has gone into it is something else.

When Gerry Marmion takes his black Ferguson TE-D20 4x4 Mk2 to any event he certainly has more people around it than any standard grey Ferguson. Congratulations to all involved in the building of this fascinating machine. ➥

Gerry undertook a great challenge and with technical know-how and determination, the project has turned out well. ■

The blue Fergie

When Derek Probert was given a hedge rescue TE-20 to use for spare parts for another restoration, he had no idea that he had taken delivery of something rather special, as Steve Wright finds out.

The 'grey Fergie', as readers will know, was not just one tractor. They may have all been produced in grey, but a very wide variation of models was produced between 1946 and 1956. Many were converted by aftermarket specialists and these variations are particularly collectable. However, despite their iconic status and unrivalled claim to fame in tractor history, some enthusiasts consider these tractors rather dull, preferring something a little more elaborate. Perhaps the TE-20s were just too common with more than half a million having been built.

Other enthusiasts recognise the importance of this tractor and also have fond memories of seeing them hard at work during the 1950s, '60s and beyond. This was the case with Derek, and the main reason for his decision to buy and restore one himself.

While in the process of carrying out this restoration, Derek's father-in-law mentioned that there was a rough hedge rescue tractor near the farm where he worked. It was thought this could prove useful for donor parts. However, as this tractor was a six-volt straight petrol engine model it was not really ideal for the particular parts he wanted at the time.

He was in the process of restoring a 12-volt petrol/tvo tractor. For this reason he was not really that interested in the offer at first. However, he decided he would have it when it was made clear that he could take it away for free!

Perhaps surprisingly, when the tractor was pulled out of its resting place all four tyres were still inflated. This made the recovery a little easier and the tractor was towed back four miles with a Range Rover.

Derek Probert aboard his ex-Brighton Corporation TE-A20.

True colours

It was not long before Derek discovered something unusual about his new Fergie. It was not a rare model or aftermarket conversion, but once it had been given a thorough steam clean it became clear that this tractor was not a 'grey' Fergie. It was blue.

It is not uncommon for these tractors to be painted many different colours by owners for one reason or another. In this case though, the wheels and the bodywork had been painted, and possibly professionally.

On a closer inspection lettering could be read along both sides of the bonnet, this area was cleaned up particularly carefully to reveal the words 'BRIGHTON CORPORATION'. At this point it was clear that this was not going to be a donor tractor, instead it would be Derek's next restoration project.

With the tractor having a clear registration plate (KCD 668), it was not too difficult to trace its history. Derek's friend Dave Noel helped here, checking the Sussex County Council records at Lewes.

The tractor, serial number 109783, was one of 85 built on Wednesday 5 October, 1949. This particular tractor was factory ordered by Brighton Council as part of the maintenance fleet working in the areas of parks, road transport and the racecourse.

TE-20 No1

While sorting through his restoration photos Derek found a picture of himself driving a 'grey Fergie'. To all intents and purposes this looked like any other TE-20 that you might see at a show. However, he pointed out that it was in fact the first TE-20 to roll off the production line. I wonder how many people have had the pleasure of driving that?

The event, at the 2006 Rudgwick Steam Show, marked the 60th anniversary of the TE-20 and Derek, along with Dave Noel, was looking after the tractor. This was not long before the Banner Lane collection of Ferguson and Massey Ferguson tractors was dispersed.

This one, along with many of the other exhibits, is now on display at the Massey Ferguson factory in Beauvais, France.

Restoration

Throughout the restoration as many original parts as possible were repaired rather than replaced. The radiator was full of acorns, so the Standard petrol engine was just briefly started to check it out.

Fortunately the engine seemed to be in relatively good condition, probably having been well maintained by Brighton Council. Very little work was required here.

The tyres, while just about holding air, were really too far gone to keep, but luckily, suitably original replacements were acquired in exchange for some spare parts which were not needed.

Everything was stripped down and assessed. It was then cleaned, repaired or replaced. The paintwork was done in Derek's garage, at the bottom of his garden, with some of the spraying being done in the ➡

After steam cleaning, the tractor was clearly showing a strong blue paint job.

The skid unit primed and ready for top coat.

Thankfully the Standard petrol engine was in pretty good condition and required little work.

The Brighton Corportaion emblem.

These tyres are hard to get as they are no longer produced, and are in good condition. The drawbar and hitch here are ready for an attachment!

garden. This was in good weather in April before too many insects were flying around!

To get the finishing touches right Derek went to visit a museum in Brighton to find the correct coat-of-arms for that year. He managed to get a colour copy to take away which was then scanned by a firm of sign writers. Four transfers were produced, one for each side and two spares.

When it came to the lettering of the 'BRIGHTON CORPORATION', the original letters were carefully traced to find out the font used and the size of the type face. Replacement decals could then be ordered.

The tractor is now finished and is a popular attraction at many local and national events. However, Derek is on the lookout for the appropriate implement to restore – one which the tractor may well have been worked with during its working days in the Brighton area. Another important consideration is the length/size of the implement for ease of transportation with a trailer. Derek has the 'BRIGHTON CORPORATION' decals ready but just needs to find the right attachment! ■

Despite its rather odd colour scheme, the TE-A20 is every inch a Ferguson tractor – compact, stylish and functional.

Ruth Jones says goodbye to the tractor she has owned and used for over 25 years.

Becoming a ploughman

Dave Beare embarks on a long-term project to become a proper ploughman. First, he has to find a tractor.

"D'you want a nice cheap tractor?" asks Morris Rogers (my neighbour from the next door farm), leaning out of his Land Rover as we slowly pass each other in our lane one day.

"Not really Morris, already got a tractor, thanks."

"It's a good one, needs moving out soon mind, it's on a farm down there (nods toward the valley)..."

"Er, well, you know, I wasn't planning on buying another tractor." (Not the usual response, but Mrs Beare is sitting next to me in the car).

Morris considers this: "No harm in having a look is there?"

"Um, well Morris, I might think about it. When does it need to be moved by?"

"Next week, like."

Oh, dear...

So that's how all this started. The tractor was indeed very nice, a 1948 (s/n 41543) Ferguson TE-A20 with tvo conversion, owned by Roger and Ruth Jones and working for them for over 25 years on Ty-Gwyn Farm.

Sadly, Roger died recently and Ruth was selling up to move, so the Ferguson had to go quickly because buyers of the property wanted to move in and Ruth was off to see her son in Australia. Next week. How could I resist my duty in buying a tractor to help a lady in distress? Impossible.

A few days later a deal was done; the Ferguson dragged from its slumbers in the barn, tyres pumped up, link-box lifted and, filled with bits and pieces and chained to the back of Morris and Hywel's Land Rover, off it went.

On to the main road and the link-box starts to drag on the ground...oops. Quick, into third gear, lift lever pulled and up she comes... good, hydraulics seem to work OK and 40psi of engine oil pressure at low revs can't be bad.

Turn off up The Pitch, a testingly-steep half-mile hill. The Landy is in low-ratio and working hard; on up another two miles of tight lanes and we're home.

First impressions are that steering is good, brakes rubbish, tin work not rusty, just a bit dented, and tyres need replacing – but otherwise the Fergie is well looked-after and un-abused.

For the next few months the tractor wasn't touched, but a cunning plan began to take shape.

Over a number of years I have reported on many ploughing matches in Wales and on occasion I have watched the Ferguson TE-20/Ford-Ferguson ploughing class in action, noticing it to be an ideal starter-class for beginners.

The cunning plan rapidly became a reality – using my 'new' Ferguson TE-A20 and an eligible Ferguson 10in two-furrow ➥

Now for a closer look!

plough, why not learn to plough properly with tuition from a number of well-known ploughmen and write the lessons and experiences up for all to enjoy?

After all, lots of tractor enthusiasts visit ploughing matches and, no doubt, many would like to give it a go but are rather daunted. So I am prepared to make a fool of myself in the interest of spreading knowledge and helping others to get started!

It is an unfortunate reality that nowadays ploughing is in rapid decline. A generation ago nearly all smaller self-sufficient farmers ploughed several fields each year, to grow arable crops and food for the farm, with their own small to medium-sized tractors. Most of them knew how to plough from knowledge passed on to them by their fathers, whereas nowadays a contractor is usually called in to plough a field.

Attending a ploughing match these days reveals many, many greying heads but very few young or female enthusiasts. Old-fashioned skills like ploughing will die away unless they are passed on; so publishing useful information on these basic skills is essential if the art of ploughing is to survive.

But first – the tractor! A more detailed inspection revealed a desperate need for new front tyres (a miracle they didn't blow on the way home) and 16in non-original rims. A pair of correct early 19in rims will be found, blasted, painted and fitted with 4.00x19 tyres, like they should be for 1948.

The rear 10-28s are sound but well worn and not ideal for ploughing, so new rubber is needed all round. Both half-shaft oil seals have failed, letting oil on the brakes (glad we

Oil leaking – a classic symptom of failed half-shaft seals resulting in non-functioning brakes.

This gasket arrangement will have to be changed if the correct compression ratio is to be obtained.

More leaking seals to be replaced!

weren't going down The Pitch instead of up – there's a busy main road at the bottom!)

The steering box arm seals are also past it and need replacing. Officially, the whole steering box has to be removed and dismantled to achieve this, but unofficially it is said that seals can be replaced without major dismantling by first grinding off the small shoulders visible at the top of the arms, disconnecting the steering joint and feeding the new seal up into place. I will keep you posted on that one.

While on the subject of oil, a quick check of the dipstick showed a full level and amazingly clean sump contents and the transmission dipstick revealed the back end to be full of clean oil too, further proof of good maintenance over many years.

There was less luck with the radiator. It looks to be well past it so, on removal, it will be sent off to Serck for a re-core, as ploughing needs a good cooling system. The water pump will be checked and all cooling hoses replaced – the ones on the tractor look like they have been there for 60 years. The thermostat will also probably need replacing; a higher-temperature rating one was usually fitted to tvo engines, enabling them to run hotter.

The Ferguson tractor was not available in a factory tvo-burning version until April 1949, largely due to resistance from Harry Ferguson, who believed vaporising oil diluted engine lubricant and so increased engine wear.

This tractor has what is probably a Vapormatic tvo conversion, which will be removed, and the engine changed back to petrol-only, as it was originally. This will also mean lifting the cylinder head to remove one of two spacers used to reduce an already-low 5.77:1 compression ratio even further for tvo burning. It will be a good opportunity to carry out a de-coke and lap in the valves, if necessary.

Turning to the back of the TE-A20, everything needs to be in tip-top condition for ploughing, but checking the lift-arms revealed a lot of wear. Various clevis pins in the lift-rods and the ball mountings of the bottom links were badly worn and will need replacing, the levelling-box on the right has seized completely and will have to come apart, the clevis in the right-hand lift rod is seized in the yoke and has not responded to the two-hammers-and-a-socket technique, so heat will need to be applied.

The check chains are badly stretched and rusted and both pivot pins connecting lift rods to bottom links are in a bad way, so all will be replaced.

On the plus side, the hydraulics seem to work well – a simple test is to pull the control lever to the top, grab hold of the pto shaft and turn it back and forth by hand; the lift arms should rise steadily.

Pump the bottom links up about a foot and then stand on the ends (careful, the links move sideways) for a minute or two. If the links stay up, the hydraulics should be OK. Checking the transmission oil (shared by the hydraulics) is at the correct level is also important.

I should say that the tractor will not be fully restored, but will be a mechanically-sound one in "working clothes" prepared for ploughing. ■

A typical Ferguson 10in two-furrow plough to be used in the Ferguson class at ploughing matches.

Ready for the Jizer and jet-wash...

Dismantling the TE-A

Dave Beare aims to be a ploughman. He has bought a Ferguson TE-A20, now he starts work on it.

Starting to dismantle a 'new' tractor project is to embark on a voyage into the unknown. You have a fair idea what should be under the accumulated grease and grot of a lifetime's hard work, but no idea of what awaits you in the innermost workings.

Often a life-time's bodging, 'temporary' repairs and abuse are only too apparent on the surface, leading to that sinking feeling about what horrors might lurk within. So, armed with boundless optimism, a copy of the factory Ferguson service manual, a load of AF spanners and sockets, a club-hammer or two, pry-bars, a 30-inch socket bar and a propane torch (to light the way....), in we go.

Fortunately, this TE-A20 had been well looked after by its last owners Roger and Ruth Jones, but it had also worked hard for over 60 years – so the prospect of wear and tear to repair was real.

The bonnet came off easily (it was surprisingly heavy), followed by a rather less co-operative fuel tank. Fixing-bolts were undone but the tank wouldn't budge, glued solidly in place by a mess of degraded rubber mounts. The pry-bars made an early appearance but were used gently in order not to bend the tank's out-turned flanges and possibly cause fuel leaks.

Next the radiator. There was no need to drain it as the cooling system was emptied many moons ago. Discovering a soldered repair, lots of damaged fins and a kilo or so of internal scale mean it will definitely be re-cored or replaced. It was time also to admire how cleverly designed and well made a Ferguson was. Even something as modest as the radiator fixings are designed with thought and field repairs in mind.

Once the hoses were off (Stanley knife – more sticky rubber) the two bottom nuts only needed unwinding a couple of turns (slotted mount holes), a top link split-pin removed, the link swung away and off the radiator came to reveal the cooling fan fitted back-to-front – no doubt deliberately in order to decrease airflow and making the engine run hotter for burning tvo.

The tvo kit was removed next. The big alloy cover over the exhaust and inlet manifolds came off easily, a load of rusty wire-wool stuffing intended to keep the heat in and assist tvo vaporisation went in the bin. The manifolds look good.

Now for the thermostat housing top; an iron casting held down to alloy with steel bolts – never a good recipe for ease of dismantling.

I undid the bolts very gently, having sprayed liberally with WD-40 and, thank goodness, neither seizes and rips the threads out of the alloy – a new casting is available but costs about 60 quid.

Lever the thermostat out and... goodness me, stuffed with Daz! When alloy castings are kept damp for a long time they corrode

Corroded alloy in the thermostat due to a drained cooling system.

... and the result of the de-grease.

and produce a flaky white/green substance, known to crashed warplane diggers (don't ask) as Daz because it resembles the washing powder. Just thought you'd like to know.

The water temperature gauge bulb is left screwed into the alloy housing as it's totally stuck in place and a test with a hot-air gun shows the gauge still works, so best to leave it in place and just be mindful of not breaking the capillary tube while working around it.

Next, the fan comes off, followed by the fan belt and water pump, and a quick twirl of the pump pulley shows up grumbling bearings – so that will need replacing, as does the belt. Various rods and linkages of the governor were also disconnected.

The front end is now exposed in full grimy glory, so the tractor is wheeled out into the yard and the filthy task of several hours scrubbing with stiff brushes and a bucket of Jizer to remove decades of accumulated oil-bound filth is undertaken, followed by blasting with the pressure-washer.

Back in the barn again, the next job is to lift the head, remove one of the two or more head gaskets and check the condition of the top end. Manifolds off first. Thankfully the nuts haven't rusted solid over the years, though a few come out complete with their studs.

The head nuts undo without problems but again a couple take their studs out. I was taught to gradually undo head nuts a quarter of a turn at a time and in the reverse sequence to tightening down the head (this instruction was probably for alloy heads), but the official Ferguson service manual makes no mention of the need to do this.

Reading section C8 of the manual, sub-heading *To Remove and Dismantle Cylinder Head Assembly*, item 2 has the fateful phrase: "Pulling at thermostat flange at forward end and rocker cover flange at rear, remove cylinder head and gasket." Oh really? Perhaps that works with new engines or if you are Arnold Schwarzenegger.

Despite much heaving, the head remains firmly attached to the block. Exploratory gentle levering lifts the front a fraction but the back stays put, the diagnosis in the end being that the rear right-hand stud is firmly stuck inside the head.

Serious levering or using wedges will damage both head and block faces so out comes the stud-extractor, having first tried running a nut down the stud and tapping it about with a hammer. ➡

Hard at work on the engine.

Stud extractors are brutal things that mangle studs, so the jammed one was scrap by the time it yielded to lots of lbs/ft torque applied with the bar. By using this method you also run the risk of shearing the stud off inside the head.

The head was then lifted free to reveal a fair amount of carbon and slight wear-ridges at the top of the bores but nothing serious. The valves came out to assess guide wear and seat condition; there was some lateral play in the guides, probably more than the .003in to .005in allowed when new, so they might be replaced.

A couple of valve seats were slightly rusty due to the valves being open for some time but these will probably tidy up with a seat-cutter. The valves and respective springs were kept in the order they came out in so they can go back in the same place, the springs will be measured for free height (should be about 1.787in) in case any are weak or have suffered from being compressed for several years.

The plough

Some people will question the wisdom of using a Ferguson plough and tractor to learn to plough when there are doubtless many better, easier-to-use ploughs and more suitable tractors on the market. However, my decision has also to do with availability and price. The Ferguson TE-20 class is affordable and although the equipment is basic, untold thousands of farmers learned basic ploughing skills using such machinery, which is what it was all about in the 1950s and still is.

There's only one place worth visiting to find a plough – the Tractormart free ads in the back pages of *Tractor & Machinery*, that's where. While thumbing through a copy of the magazine, I noticed a small advert for a Ferguson 10in plough in restored condition. A quick phone call to advertiser Bill Heath revealed it was just what I was looking for and so a deal was done. Bill said he could have sold the plough ten times over since the first phone call, that's how effective the Tractormart section is!

Good used tyres from Hubert Thomas.

The plough was a Ferguson 10HC-AE-A28 and was in excellent condition. Just a couple of minor repairs were needed – the rear coulter (disc) stem seat needed building up with a little weld in order for the disc to run vertically and a couple of new shares needed some grinding off so the mounting holes would line up.

A couple of vital things to buy when embarking on learning to plough – copies of the *Ferguson Mouldboard Plough Manual* and Brian Bell's excellent book *The Tractor Ploughing Manual.*

Regulations for the Ferguson TE-20 and Ford-Ferguson match-ploughing class allow the use of a standard tractor and 10in two-furrow Ferguson mouldboard plough with standard Ferguson top-link. Full regulations are published by Robert Godfrey and can be obtained from him at Park Lodge, Rhosddu Road, Wrexham LL11 1NF, tel: 01978 367000, e-mail: robertgodfrey@goded.co.uk or from the website: www.ferguson20rules.co.uk

It is hoped that the Ferguson TE-20 class will be included in the British National Ploughing Championships at some point in the future, which should make this class more popular and competitive. ■

Getting the compression ratio right for tvo requires the use of the correct number of gaskets/spacers.

Garfield Andrew adjusts the plough.

Understanding the plough

Dave Beare prepares his Ferguson tractor and plough combination and digs it in for the first time.

Work on the Ferguson TE-20 has by force of circumstances been limited to the engine top end – not a very exciting job but nonetheless essential if you want your Ferguson to run well.

The head came off to remove all the gaskets and spacers fitted to lower the compression ratio for running on tvo, the plan being to return the tractor to original 1948 petrol-only spec.

As the head was off it made sense to carry out a valve gear check-over once the nasty scaly, oily mess found all over it was scraped and washed clean.

The rocker shaft was already off and the pushrods were retrieved from their holes and lined up in the order they came out. Probably not necessary as the valves are going to need work anyway, but I prefer to do it.

While on the subject of pushrods, roll them along a flat surface to check for straightness (a bent one will immediately show up) and, when cleaned off, wipe them over with an oily rag to deter rust from forming.

Next, the filth-encrusted rocker-shaft; dunk it in a trough of solvent (paraffin or Jizer, not petrol) and leave overnight to marinade well, brush-clean off and blow dry with an air-line.

Drive out the pin holding an end-cap on and remove end-cap plus everything on the rocker shaft – keeping it all in order as it comes off – then clean everything again, using a scratch wire-brush to finish off.

There's a lot of stress and impact loading on valve-gear and lubrication is a bit sketchy, so appreciable wear can take place, which will affect engine efficiency.

Cleaning the rockers (they are handed so make sure they go back where they came from) includes scraping the crud out of the oil groove in the centre of the bronze bush with an angled scribe (no, not a bent journalist) and poking a piece of wire through the oil-hole that supplies the pushrod top to clear it. ➥

Drive the locating pin from the rocker-shaft end-cap and remove rockers etc, keeping everything in order.

Checking valve spring free length; all were within spec.

Then wash each item off again – carbonised oil deposits are very abrasive.

Remove the second end-cap from the rocker shaft, run a wire with a small piece of rag through it and blast out all the holes with compressed air as the hollow shaft harbours lots of grot.

Check the shaft for wear – this one is well-worn by the rockers, the reason probably being that the very small oil-supply drilling through the rear pedestal was almost blocked with debris, cutting off the oil supply so the rockers were running virtually dry.

Problem: new rocker shafts are unobtainable at present, so it will have to go back in again until new ones are available.

Moving on to the valves and seats, three out of four exhaust seats were very badly pitted and the others pretty worn (no hardened inserts) so all eight were re-cut and the valves lapped in with coarse and fine paste, using a good old-fashioned stick and sucker.

A nice little discovery – possibly when a de-coke was done, someone number-punched the valve heads 1 to 8 and the exhaust valves were punched 'EX' as well (they are made of a different metal to the inlets).

The valves, springs, caps and collets were then refitted and a little wipe of oil on each seat and valve stem helped keep it all lubricated and rust-free, ready for start-up time.

The plough

The ex-Bill Heath 10HC-AE-A28 two-furrow 10in Ferguson plough dates back to 1949, so is almost the same age at the 1948 TE-20 tractor that will eventually be carrying it.

A close inspection revealed it to be in very good restored condition (thanks again Bill), both beams straight and true, shiny un-pitted mouldboards, discs re-bushed, the furrow wheel bearing good and wheel spring intact. Only a couple of new shares (still available) were needed to ready the plough for a trial run.

The plough was attached to my FE-35 and taken off down the lane to Garfield Andrew's farm, a neighbour who still competes regularly in matches.

He offered early on to give me some tuition and help set up the plough, saying "I've got this practice plot, on the hill and away from the lane – nobody will see you making a b****y mess up there!" How can you refuse a generous offer like that?

I'm still learning the names of all the different parts of a plough and what they do but just wanted to get it in the ground and see how it worked, so one early winter afternoon we had a go on Garfield's secret plot.

First job was to check the tractor was set up for working with a Ferguson 10in two-furrow plough – measuring the width between the inside faces of the rear tyres to ensure it was 40in (two widths of the two-furrow 10in plough) and doing the same at the front, also checking the front and rear right-hand inside tyre faces will run in line and are parallel.

Next, the two coulter discs needed to be adjusted in height and distance in relation

Lapping the valves the old-fashioned way.

Ferguson plough mounted on the FE-35.

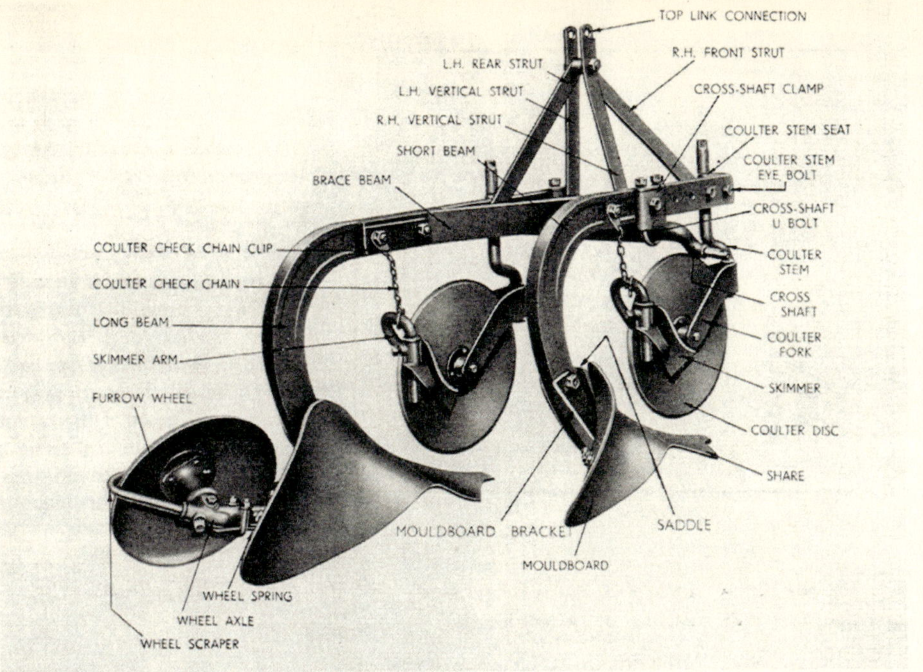

Above: Ferguson plough component names that are well worth remembering when entering the world of competition ploughing.

A little fine tuning was required to straighten up the rear coulter so that it runs true and vertical.

Right: The first few furrows of working with the Ferguson FE-35 and 10HC-AE-A28 plough – it works!

to the fixed shares. They are mounted on a cranked coulter stem and clamped by a large eyebolt.

A basic 'starter' adjustment is 1¾in between the bottom of the discs and top of the shares, with the disc lifted by hand to take up any slack in the coulter mounting forks. The distance between the side of the coulters and landside of the shares should be about ½in to start and is achieved by turning the coulter stem.

The final adjustment was to set the cranked cross-shaft (which attaches the plough to the tractor's lower links and also alters the width of the front furrow – the rear furrow width is fixed) vertically, the right side crank pointing down, the left side up.

We decided to run along the fence on an unimproved old pasture, the ground was nice and firm and would cut well (very little rain in the autumn meant an arable plot would be too dry for a try-out) so we dropped the plough and dug it in.

Amazingly enough it ran almost straight and fairly central, so the next step was to stop and adjust the top link until the plough sat level in the ground, then use the hydraulic lift to adjust the furrow depth to around 6in, then turn the cross-shaft until the front furrow was cutting a measurable 10in from the right-hand inner rear tyre wall and finally work out what setting of the levelling-box on the right-hand lift link gave two reasonably equal furrows.

All these minor adjustments, together with running on a bit to test how the plough was responding to each change, took up most of the afternoon, but we finally got a couple of sort-of straight, consistent depth and equal sized furrows! Success! ■

In at the deep end

Our trainee ploughman, Dave Beare, is thrust into competition.

The phone rings one dark and blustery evening.

"Dave, John 'Ooson 'ere." (John Hooson is one of Flintshire Vintage and Classic Tractor Society's kingpins, very active in ploughing matches, road runs and always up for a laugh).

"Hello John, how are you, what's happening?"

"Listen Dave, what you doin' on Sunday 2 December?

"Er, um, don't really know yet, why?"

"Fancy 'avin' a go at a ploughing match?"

"Blimey John, I don't know what half the parts on a plough are called, never mind being able to use it!"

"Doesn't matter mate, it's just for fun and to say Happy 70th Birthday to Ian Wilson, but it's a big secret, he don't know about it, so don't you tell anyone."

"Right. Well, er, OK... but I've no Idea how to plough, only cut three furrows in me life, so I'll be making a complete pig's ear of it, and the TE-20's not running yet..."

"Don't worry, nobody's gonna mind. I've asked Sonny Smith if he'd give you a bit of a hand, show you the basics, like."

"Good heavens, I'm sure Sonny's got better things to do on a cold December morning than stand ankle deep in mud, watching a right old mess!"

Anyway, the upshot of our conversation was that I was signed up for a first-ever match, organised in great secrecy by John Hooson and David Hankey, the Ian Wilson 70th Birthday Celebration Ploughing Match.

Ian was told it was a practice day for some local ploughmen and he had to come along. Ian's response was something along the lines of "John, 'ave you completely lost it? A practice day early December! You've got to be crazy." He nevertheless agreed to attend, despite serious misgivings about John's sanity.

I'd so far only spent a few hours with neighbouring 'tutor', Garfield Andrew, trying out basic adjustments to my Ferguson plough, never mind trying it out with many of Britain's finest ploughmen around.

The entry list was a bit like a "Who's who" of vintage ploughing... Ian Wilson of course, and Richard Ingram, John Milnes, Mike Watkins, Mike Holloway, Stuart Burden, Richard Wilson, Graham Jackson and many, many more well-known local and national ploughmen – over 50 in total. Competitors promised to come from far and wide: Crewe, Wigan, Rotherham, Clitheroe, Preston, Ripon and Warrington, all to help Ian Wilson celebrate his 70th birthday.

Despite truly awful weather with torrents of rain for two days, nearly everybody turned up on the day at Borras Head Farm, Borras, near Wrexham. What a quagmire!

Part 4

Co-conspirators David Hankey (left) and John Hooson (right) flank birthday boy Ian Wilson.

Left: The second run – taking notes of all the plough settings, both plough bodies now in the ground.

Modern 4x4 tractors were used to tow lorries and trailers with tractors on to the parking area for unloading… a strong blustery wind threw heavy showers at us, interspersed with sudden, glorious bursts of sunshine.

Just getting to the ploughing area involved slithering through a thick brown soup of mud, but we all made it and I was shown my plot, number one, right in front of the burger van. A guaranteed audience all day… thanks John!

Not only had Sonny Smith come along but George Allwood, a Society of Ploughmen member, also joined in to help with advice, so we started from the very beginning of everything.

The basic plough settings were unchanged since Garfield Andrew and I tried them back in the late autumn, but we went through them all again as this time Sonny and George were teaching an opening split and settings would be very different. Out came the sighting poles, three allowed, one placed right at the start of the plot, the second about two-thirds of the way down and the last just beyond the far edge of the plot – all exactly in line so the driver can only see one pole. You are allowed an assistant to help line up and then remove the poles as the driver approaches them.

We set up the Ferguson plough for the opening split as follows; front body lifted clear of the ground, using the right lift-rod levelling box wound up to the top and, if necessary, lengthening the top link. The rear body alone does the work and a shallow 2-3inch deep furrow is what's needed, set with (in the case of a Ferguson tractor and plough) the hydraulic lift.

Slow and careful driving is vital at this stage as all the subsequent furrow alignments depend on this first one; the centre of the bonnet should be lined up exactly with the poles – a sighting dollop of soil stuck to the front (or a rubber rabbit in the case of Tony Bradley) to help keep you straight. And remember to maintain the same driving position!

Most experienced ploughmen crouch dead centre, noses above the steering wheel boss, sighting down the bonnet. If you want to check the plough, always stop first before turning round; if you turn your body while steering you will inevitably pull the steering off a straight line.

At the end of the opening split keep going straight as the marker cross-cut furrow is crossed, lifting the plough once clear. It is worth making a note of the top link length pin to pin as this setting, once established, can be used in the future. Then turn the tractor round and line up for the second run. The plough settings are different this time as the aim is to plough two dead straight furrows, so the front body is lowered by unwinding the levelling box to level before setting off.

The front disc must "run alongside the opening furrow wall, leaving no land unploughed in the centre of the opening split" (Brian Bell). The tractor straddles the first single furrow so a different means of aiming the tractor is used, in the case of the FE-35, the side of the bonnet lines up well with the cut side.

Stop after ploughing a short distance to check your work, making any adjustments necessary, ensuring the tractor is stationary before looking back, as otherwise the furrow will not be straight.

"Making any adjustments necessary" is much easier said than done – in the case of a Ferguson 10-inch two-furrow plough only a limited number of adjustments are possible, making it perhaps easier to set up than match ploughs, which have many more. ➡

Just trying to plough straight and with consistent height furrows, not yet moved to the other side of the plot.

Correct skimmer setting; turning a small slice of soil into the furrow bottom.

Lining up to plough along the side of my neighbour's plot.

The hydraulic lift lever controls depth of the furrows, so measure the furrow depths with a tape measure and note the position of the lever when depth is correct. The skimmers are not yet in operation as a full-depth furrow has not been ploughed, this adjustment will be dealt with shortly. Consistent height of the two shallow furrows is achieved by using the levelling-box; at this stage it has to be carried out by eye.

Once the second run is completed in a match, a ploughman would wait while the judges mark his or her work. In this case I wasn't competing, so we carried on and for the third run the plough was dropped down to almost full working depth of six inches to cut two full furrows and find out how that worked.

The skimmers came into action and needed adjusting – their function is to peel small sections of soil off the face of the coulter discs and turn them into the bottom of the furrows, best illustrated by looking at the black and white photograph from the *Ferguson Ploughing Manual*. The points of the skimmers should lightly touch the discs and height adjustment should be such that the points penetrate about an inch into the soil.

After a couple more runs under supervision, Sonny and George left me to carry on and went off for some well-deserved lunch, so I played around for part of the afternoon, working out which adjustment had what effect and learning one cardinal rule. Adjust one thing at a time, try it out, if it works, good, if not, go back to where you were before that one adjustment before trying another. Change two or three things together and soon you have no idea which adjustment altered what and are lost. One at a time, one at a time…

With many thanks to Sonny Smith and George Allwood for the expert tuition, Jackie Smith for taking the photos, Mike Holloway for tractor transportation and John Hooson and David Hankey for a great, if rather nerve-wracking day out! ■

I didn't finish, but I'm fairly pleased with the first attempt.

Getting it running

Dave Beare prepares his Fergie for work – there's a lot to do.

The poor old Ferguson TE-A20 that I am supposed to be using to plough with has languished in the barn in part-dismantled state, with many bits being slowly reconditioned as and when time and temperature permits. Finally it has stopped freezing and some free time has arrived so it is now full-steam-ahead.

You may recall that this 1948 TE-A20 was a petrol tractor converted to tvo with a Vapormatic kit some time in the early 1950s and that I am converting it back to straight petrol, as it was when it was made at Banner Lane.

Good idea in theory, but a number of complications arise.

The 1950s conversion raised the head relative to the block (reducing the compression ratio for burning tvo) and, as a result, the original push-rods were then too short. Vapormatic supplied a set of longer push-rods in the conversion kit to compensate for this but, in reverting to a normal thin gasket, the head comes down closer to the block. The push-rods are now too long, the rockers at the wrong angle, the valve springs in danger of coil-binding and valve adjusting screws useless. What to do?

A set of shorter standard-length push-rods could not be found (diesel TE-F20 ones won't fit – too short, we tried) so either the push-rods needed to be shortened or the rocker shaft pedestals had to be shimmed to raise the whole assembly. After some careful measuring and consultations with a couple of local agricultural engineers, we decided the latter course was more practical.

Initial set-up was carried out using a pile of 1½in-diameter penny washers, checked for consistent thickness, and eventually four per rocker pedestal were found to give exactly the right height, closed rockers were level and valve clearances could be adjusted correctly (0.010in inlet, 0.012in exhaust).

The engine was turned over by hand to check clearances and it all looked fine, then the shaft was removed and an oil feed hole drilled in the rear set of washers as the rocker shaft and rockers are pressure-lubricated via a passage in the head and the rear pedestal. They were then lightly centre-punched and sealed with a smear of gasket sealant between them to prevent rotation or leaking. ➥

The plan is to have four single-piece steel shims profile-cut to the same shape as the base of the pedestals but, for the time being, the washers will just have to do.

Next complication was with the carburettor – a Zenith 24T-2 updraft instrument common to both petrol and tvo (and lamp-oil) TE-A20s but with some differences according to fuel. A larger main jet is fitted for tvo operation (1.05mm for tvo, 1.00mm for petrol) and an extra slow-running fuel feed drilling is made.

The larger tvo main jet can either be replaced with a 1.00mm (if you can find one) or compensated for by the main jet adjuster screw at the base of the carb, the extra slow-running drilling (see diagram – X marks the drilling) could be blocked by removing a small lead plug at the base of the carb to access it, but I'm going to run the tractor without doing this to find out if it's essential. Quite possibly not.

The major problem with this carb was the tremendous amount of wear at the throttle butterfly spindle, mostly in the zinc-cast body but some on the brass spindle, which would allow air to leak past and affect both mixture and idle settings.

Some carburettor specialists offer a rebushing service for spindles (Burlen Fuel Systems may be able to help), but I managed to use the top off a similar carb from a car with much less wear to rectify the problem.

Cleaning out the carb was a lengthy business – internally it was encrusted with filth and deposits; be very careful when dismantling as brass jets and small screws can be corroded into place – use the correct size screwdriver as it is easy to rip the slots apart and then you are really snookered. Likewise, when removing the tiny screws holding the butterfly and spindle together – these have split ends to prevent them unscrewing in use.

Checking no valve-spring coil-binding takes place.

Next was the oil filter problem. This TE-A20 has the early vertical filter and, for the sake of originality, I'd like to keep it (the vertical filter assembly can be replaced with the later inclined assembly) but it seems impossible to find replacement filter elements – they are of corrugated felt.

The solution was to spend a happy hour or so in the local tractor spares department trial-fitting different metal/paper elements that *are* readily available. The closest was a Bepco 96-2, which fits a multitude of Ford and MF tractors. The length is right but the top hole has to be enlarged to fit the vertical filter spigot (carried out with a rotary cutter – clean the swarf out very carefully), the bottom hole fits tightly on the conical boss inside the filter base.

The front end was going back together well. A new lower temperature thermostat was fitted to replace the hotter-running 75° tvo unit, the new water pump and rebuilt radiator were put in place with all-new hoses and clips; reconditioned carb and manifolds likewise.

Removal of the inspection plate on the right of the sump revealed a quantity of thick black sludge in the bottom; it was scraped out and then brushed clean with paraffin, making sure it was dry in there before replacing the inspection plate and mesh screen.

Moving to the back of the tractor, off came the mudguards, eventually; they had been replaced in the past and somebody used

Checking butterfly spindle wear – appreciable up/down play needs to be rectified.

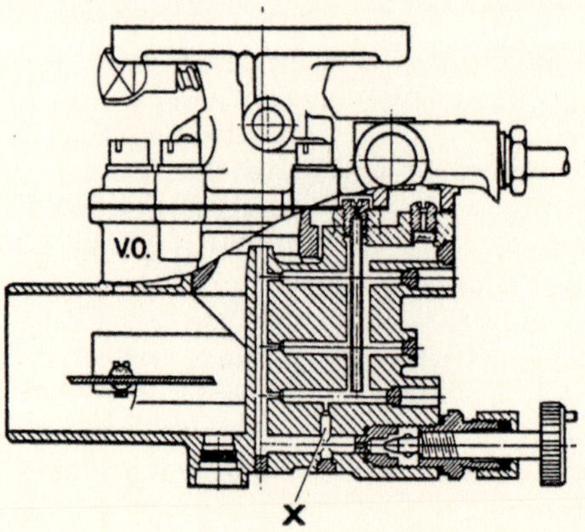

Extra slow-running drilling for tvo at X.

The felt oil filter element and vertical assembly.

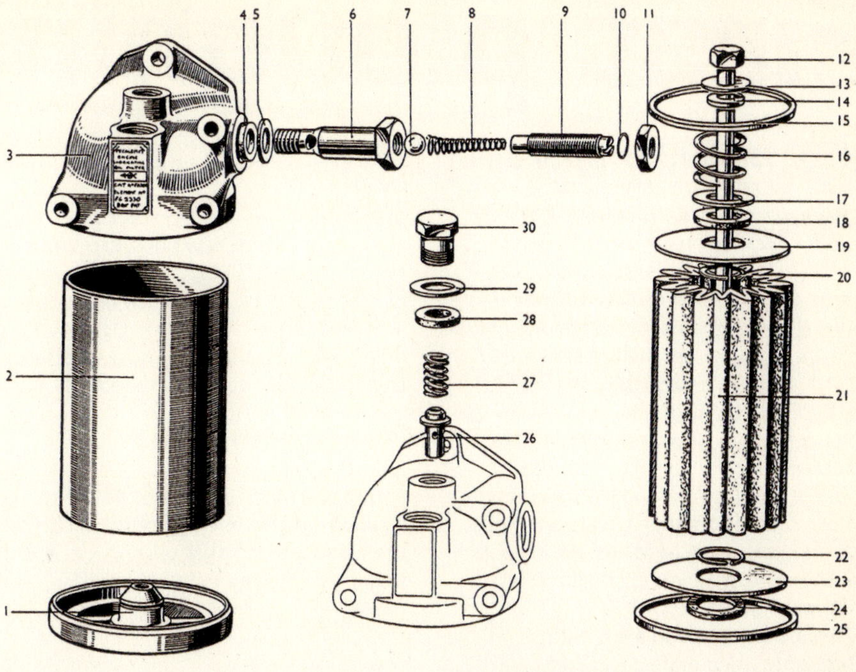

Internal components of the vertical oil filter assembly.

round-headed coach-bolts, all now nicely rusted in place… an angle-grinder job in the end. The wheels came off easily, as did the brake drums to reveal thoroughly oil-soaked brakes. Six studs and nuts hold the half-shaft, bearing, seal and housing in place: on removing this a pint or so of oil gushed out – possibly indicating that the back end had been overfilled.

The outer hub seals are likely to be past it but removing them is a nightmare; a shrunk-on collar behind the bearing on the half-shaft has to be drilled and broken off and the bearing itself often takes a 10-ton press to shift before you get to the seal. One local agricultural engineer advises using easily-replaced later MF inner half-shaft seals to keep the oil in, not over-filling the back end and packing the bearings with grease, which should stay put with the old outer seals still in place.

Relining the brakes is next, resetting the linkages, a coat of paint on the back end and refitting of a hopefully revitalised six-volt starter that might just throw the Bendix into engagement with the flywheel and allow me to start the engine for the very first time!

Working on a little grey Fergie is a great pleasure – it's a properly-engineered machine full of good practical ideas, designed and manufactured by engineers who took real pride in what they made out of metal – there's almost no sign of plastic! It all fits together superbly well – a real classic of form and function, no wonder it started a revolution and was a world-beater. ■

Checking the diameter and fitting the made-up paper gasket.

Above: Oil, oil everywhere…

Right: Out comes the half-shaft assembly, off come the brake back-plate and linkage.

First ploughing match

After all the preparation Dave Beare finally enters his first match.

Work on the TE-A20 has proceeded with increased urgency as there was a deadline to meet – match ploughing for the first time using both this tractor and the 10HC-AE-A28 plough.

My debut was to be at the Offa's Dyke Ploughing Society's event on Sunday 20 April but by early April the tractor was still on axle-stands and hadn't yet run, the oil-soaked brakes were all in bits, the tin work was in a shed, the plough and links still on the back of the FE-35…

First job was to clean the brakes thoroughly, using old tvo – it works well – then drill out the brake-lining rivets, discard the old linings (asbestos-based, so disposed of in an asbestos recycling container) and replace them with new ones. Remember, modern resin-based linings don't bend like old asbestos ones, they crack very easily.

Everything was dismantled, cleaned and dried, the brass shoe-adjusters stripped, cleaned and lubricated with red brake grease, *not* the copper-based stuff which dries out, goes solid and will seize the adjusters in time.

Hub and half-shaft assemblies were cleaned – stuff some rag round the inside of the bearing to stop grit getting in – and the bearing tracks and rollers checked for signs of blueing or damage; thankfully they were still serviceable.

Uncle Phil's marvellous new inner oil seals were tapped into place (having first greased the wheel bearings). The seals were old Payen stock, part number NA622, but new alternatives should be obtainable from any good bearing and seal distributor. Dimensions are; 2⅞ x 3.762 x ⅜in (73 x 95.56 x 9.5mm). The backplates went back on, in went the half-shafts, hubs tightened, end-float checked and adjusted by shims fitted behind the backplates. Brake shoes and adjusters went on and the actuating cams

First sortie, no wings or bonnet, but it runs!

went back in place (this being a June 1948 tractor it has those peculiar offset "kidney cams" that are fitted flat-side up), then various springs and things replaced and the drums tapped on.

Constant referral to the *Ferguson TE-20 Service Manual* illustrations is essential – it would be easy to fit bits the wrong way round, not having noticed which way up they came out and only discover the brakes don't work when the wheels are back on.

Final adjustment is made by slackening off two anchor-pin nuts either side of the brake shaft, fully expanding the shoes, tapping the nuts to settle the anchor pins in their new position, expanding the shoes again, then tightening the nuts up hard. For now, the old wheels and tyres will just have to go back on again, time being short to fit the good tyres and rims to centres.

Meanwhile Gordon the Electric had checked the 6V starter and pronounced it fit.

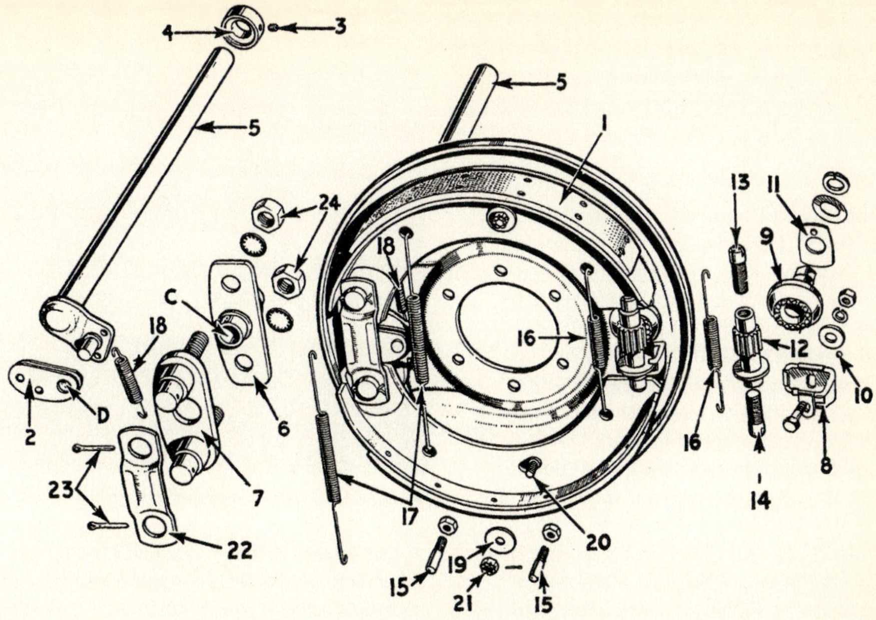

FIG. 2
KIDNEY CAM OPERATED BRAKE

Kidney-cam brakes are quite complex and need careful adjustment.

"And why won't it throw the Bendix into the flywheel then?" I asked.

"Because I bet it's the earth connection on the dash that's stopping it and the contacts in the starter switch are burned" Gordon replied, and he was right.

Six-volt systems need bigger cables and better earthing than 12V, so the old earth lead was replaced with a heavier-gauge one, connected to a nut on the bell housing (cleaned back to bright metal and smeared with Vaseline), the starter switch reconditioned and bingo, the starter engaged first try.

Right! On goes the petrol tank, in go a gallon or two of petrol, fill up the radiator with 50/50 water-antifreeze, fill the sump with fresh 15-40 oil, check for external leaks all round and pray the liners hadn't shifted while the head was off, letting water into the sump. Never turn a headless engine unless all liners are firmly clamped in place.

Turn on the ignition and petrol tap, nip the odd leaky union up, pull out choke knob and engage starter. A bit of churning, one or two faltering coughs and, oh dear, petrol pours from the lower air-cleaner hose. The light spring holding the choke butterfly closed had jammed and the engine was sucking neat petrol. Rectify. Dry out plugs and try again.

This time much more positive signs – one or two cylinders fire and kick the starter out of engagement; another try and with a mighty roar the Standard Motor Company's finest burst into life, covering everything around it with soot.

The ignition was a tad retarded and I'd forgotten to tighten the oil filter bolt, so oil was oozing out round the top, but there was 40psi oil pressure and it also looked like the Bepco filter cartridge adaptation was working. Likewise the spacers underneath the rocker shaft pedestals, the valve-gear giving out a well-oiled hundred-grannies-knitting sound.

The tractor was insured and even had a tax disc so, once running temperature was on the clock and the engine ran smoothly, I backed it out of the barn (almost exactly a year since it went in) and went off up the lane – without tin work. Brilliant! ➥

Plumbing side of the back-to-petrol-only engine.

It runs rather rich on petrol, so the correct smaller main jet will be substituted (if I can find one) and the extra slow-running drilling blocked off, but it went well.

A couple of discoveries – the clutch is very fierce, in or out, not much in between (probably indicating it's oil-contaminated) and the brakes are excellent – something of a revelation.

Then I ran out of petrol half a mile from home and took the carb apart, thinking the main jet had blocked. Duh!

Time was very short before the Offa's Dyke Ploughing Match – just a few days left and lots of other commitments – so the old bonnet and mudguards were just bolted back on and, at tea-time the day before the match, the TE-A20 was fitted with the links, levelling box and the plough from the FE-35.

At this late stage I didn't even know if the draft control worked, but a quick run down to Mike Holloway's turnip field for a brief try-out proved it did. Another problem immediately surfaced. The tractor wouldn't plough straight. I discovered the front axle was set to 52in and it needed to be narrowed down to 48in – so at 6pm I was struggling with rusted-up bolts, trying to fix this.

Match day dawned grey, cold and wet but 45 ploughmen from all over Britain had turned out to plough, and the TE-A20 joined them. I made an appallingly bad first run, not remotely straight and had no idea what adjustments to make to the top link, levelling box and cross-shaft as the tractor and plough had never worked together before.

George Allwood wonders what's going wrong with the second run.

The second run was not much better so I asked George Allwood for some advice – he recommended starting again with the poles, so that's what I did and gradually found the right settings for the plough. The fierce clutch didn't help at all with 'ins', lifting the front wheels even at low revs, but the engine slogged away well and the governor did its job, as did the draft control which constantly worked away keeping the plough at an even depth.

Each run got a little better and work of a more acceptable standard was produced, though still some way off the quality shown by more experienced ploughmen – and there were many of them at this match.

So, what's it like as a novice to plough at a match? Very challenging but immensely rewarding when you get it even half-right; frustrating when, despite trying different adjustments, nothing goes quite right and daunting when struggling with a lack of experience while being watched by an audience of knowledgeable farmers and ploughing enthusiasts.

Will I continue to plough? Yes, most definitely. Learning to plough is very enjoyable, you will meet many sympathetic ploughmen and judges who are more than willing to help with advice and tips.

Your tractor is doing what it was built to do and even a complete novice like me can, with a little practice, produce adequate work, though to reach the standards seen at most matches will take a good few years.

I'm hooked and will continue to compete in the Ferguson class, when this is included in matches, and at local events in the Vintage Mounted class. Go on, you've got a good tractor, give ploughing a go… ■

Correcting early errors, much straighter and furrows more consistent.

Photos courtesy of Graham Porter & Sonny Smith.

The green Fergie

Dave Noel tells the story of his 1954 Ferguson TE-A20 industrial tractor, which spent its working life near Dover.

It all started in March 1995 when I went to a dealer friend in Rye, East Sussex. I was picking up spares for my 1958 Massey Ferguson FE-35, which I was in the middle of restoring. Parked in his yard was a very sad looking TE-A20 and I could see that under the rust it was green.

The fleet owner's name had been painted over with a darker green colour and with my friend's permission and some emery cloth, I rubbed it down to try to identify the original owner. The word 'Kent' was revealed and not being able to resist such an unusual find I struck a deal, borrowed a trailer, and loaded it up under its own power.

I took it to Dover to store in my mother's empty garage. When I went to unload it from the trailer it wouldn't start. This was due, it was thought, to heavy rain getting in the electrics, yet it had a flat battery (mainly due to a loose terminal) and the dynamo was not charging. I had to unload it with the winch and push it into the garage. It was locked up and, as I drove back home to Sussex, I was not sure if I had done the right thing in buying it.

A month or so later my wife, Sandra, and I spent the day at mum's, during which time I pushed the tractor out and made a long list of bits missing, and bits that were beyond repair. The list included mudguards, front grilles, front toolbox, headlamp and more. The sidelight had broken off and the rear lamp bracket on the offside had fallen off due to the lack of metal in the mudguards.

Turning to the exhaust silencer I noted it had come from a motorcycle. By the look of things, particularly the bracket on the rear axle, it originally had a down-swept exhaust system. A new wiring loom was going to be in order as the one fitted was rather moth-eaten in places and the horn switch was broken. I also took a note of the vehicle number – 390709 – as no log book came with the tractor.

Later, I contacted Friends of Ferguson Heritage in Coventry and was informed that the tractor was completed on Saturday 22 May 1954 along with 125 others. Having the date of construction I went to the Kent County Council library at Queen Elizabeth Street, Maidstone and spent a couple of hours going through the vehicle registration books.

I found the page telling me the tractor was first licensed on 11 June 1954 as SKT 934 and was one of about 24 allocated to the KCC Parks and Roads Dept. I obtained a copy of the document and sent it to the DVLA at Swansea. ➡

Detail was the key to success with the restoration of Dave Noel's 1954 Ferguson TE-A20.

Even the headlight is a genuine original example and really makes the tractor stand out amongst the crowd.

I also contacted the Head of the Roads Dept at Kent County Council, Maidstone. The manager was very helpful but had no records, he told me that it was withdrawn in the early 1970s and that all tractors were yellow by 1973. Very interestingly, he told me that due to Council policy, most KCC tractors were scrapped; only the odd one escaped dumping if it was purchased by a member of staff.

My friend in Rye informed me that he understood it had been kept in a barn for about 12-15 years to cut logs until the bearing and oil seal on the rear shaft had both disintegrated.

I left it in the garage for nearly 18 months while I completed the FE-35 then took my trailer and a new battery to Dover to collect it. I poured the petrol in the tank, connected the battery, replaced the piece of string that opened the choke, turned the key and it started. Much to my surprise the tyres had stayed up.

The Council had converted the tractors to ignition start, the same as their vans, and had disconnected the gear lever starter to avoid confusion, for the drivers who were not accustomed to Ferguson individuality.

Once home, all the tin work was removed and new skins were placed on the old mudguard frames. I rubbed the bonnet down to uncover all the signwriting. Tony Funnell from the Bluebell Railway traced and measured it all ready to carry out the job when it was nearly completed.

I stripped the paintwork down to bare metal and gave it to another associate, Ray Woolford. He straightened it out, rebuilt the front grille and repaired other items where necessary and, as always, did a fine job.

The cylinder head was removed for a de-coke and I took the opportunity to check the pistons and valves. I replaced two broken valve springs; two sheared rocker bolts and replaced non-functioning valve adjuster screws. The dipstick showed 'full', but as hardly any oil ran out of the drain plug, I took the sump off. I cleaned the sump thoroughly, inside and out, before refilling with new oil.

The gearbox selector was rebuilt and I refilled with oil, after a good clean out, as condensation was a problem in this area. I also rewired the tractor. Rear hub oil seals went on and the brakes were relined as they were contaminated with oil. The brake rods had been re-bushed, but when I tried to adjust the brakes, one side would not adjust and I found the back plate was cracked. Unfortunately, the split only showed when the brakes were applied.

After cleaning back to bare metal the fuel tank was found to be perforated, so that went off to friends at Sendmarch Tractors for repair. The radiator was sent to a local firm who repaired the header tank. Wheels were stripped and the rear rims were found to be rusted right through. On one side, the rim bead had actually collapsed.

The original rear tyres were in a fair condition, new wheels were acquired and the tyres were taken off the old rims with the help of an MF 135 to break the bead! After the first tyre had been changed and pumped up it was left by the tractor. After ten minutes there was a loud bang, the tyre had split in two, resulting in the purchase of a new pair of tyres.

Both the starter motor and dynamo were sent to CCM of Burgess Hill, Sussex for a complete rebuild as both rattled with everything inside either loose, or worn out. After many evenings in the garage with a wire brush and paint stripper the back axle, gearbox and engine were washed

off and ready for painting with the highly-recommended Bondaprimer and undercoat.

On the following day, and despite the cooling system being pressure tested, a two-inch horizontal crack was discovered on the engine block, starting from the drain tap. Mark Fletcher, a local agricultural engineer was put on the case and he thought he could repair it by welding, but would need the engine on its side. So we had to split the engine from the gearbox.

This turned out to be a blessing in disguise as we found the clutch to be on its last legs. Both the rear crankshaft oil seal and gearbox seals were leaking the fresh oil that I had poured in. Another problem was cured with a new Bendix and spring for the starter. While the engine was away for welding most of the other bits got painted so with the return of the engine it began to look a bit like a tractor again.

In the summer of 1997 I took my Ferguson FE-35 to the Great Dorset Steam Fair and met two gentlemen who had worked for Kent County Council. They gave me a lot of facts, including why the rear tyres on my TE-A20 were larger than the normal 10-28 size. It was to give extra speed between jobs.

The fleet no T66 meant T for tractor; the first 6 was for Lydden Hill depot, the second 6 denoted it was the sixth tractor in the fleet. They also told me what the tractor was used for, and that a Twose verge plough, or trailer and Bomford mid-mounted mower could be fitted. The Ferguson blade on the front was for clearing rubbish and also the snow when it fell around the Dover area.

During the next year I acquired the verge plough from Fergiland, the mower from Pembrokeshire and the blade and some of the equipment from Cheshire. Many parts of the blade assembly needed replacing and Harold Beer in Devon came to my rescue – I gave him the measurements of the missing parts and after he had made replacements the end result was very pleasing.

During 1998 the tractor made its first public appearance at the Royal Show at Stoneleigh. Unfortunately, it disgraced itself by not running very well. The coil was at fault and after it was replaced the tractor was fine at various other shows; Ardingly, Bluebell and Bromley Farm & Garden Festival. It was also on static display at the National Ploughing Match on the Friends of Ferguson Heritage stand.

The 1999 season started at Easter, taking the tractor to Scotland for the National

The carburettor required attention from George McAleer several times before the engine ran well.

Road Run which proved to be an excellent, although rather cold, weekend.

When it went to the Kent Show at Detling, I won a shield and met a lot of interesting people who had worked with the tractor, including the chap who did the original signwriting. This had all been re-created by Tony Funnell. I won a cup at the Bromley Pageant and another at the Cranleigh Agricultural Show.

The season was topped at Dorset, where on the Sunday it would not run well but the club's resident engineering genius George McAleer cured it with a tune up and we won the trophy for the Best Tractor at Show.

The last show was the Concours Show at Newark which was great fun; Cliff Yeoman won this event so I'll have to try again. The great thing about these events is meeting people from all areas and backgrounds, talking tractors and being there with lots of good friends.

In the winter of 1999 the carburettor had to be rebuilt as air was getting in the wrong places. The repair was carried out by Southern Carburettors of Wimbledon.

During 2000 I took the tractor to more shows, including the pageant at Danson Park, Ardingly, Cranleigh, Rudgwick and Dorset, finishing the season at Harrogate where on the Monday morning when loading up I noticed a water leak from the head gasket. Oh dear!

Returning to Sussex, Derek Probert, my great helper on all things mechanical, removed the head to find a lot of quite serious cracks in the block, so changed the gasket, tightened it up and hoped that it would keep going for the season, which it did.

Since then a replacement engine has been acquired and after a strip down and a thorough rebuild, it's back in the 'green beauty.' Today, the tractor comes out occasionally, when time permits. It can also be a break from running the area FoFH stand, with Sandra, at a good number of shows.

The tractor has received another repaint along the way and is very much part of the family, having been with us for some 15 years now. ■

Left: Considering this has all been rebuilt in a small garage, it just shows what can be done with limited facilities.

It's grey in Sweden

We look at a nice Ferguson TE-A20, used for a variety of jobs, in Sweden.

The owner of this Ferguson TE-A20, 21-year old mechanical engineering student Håkan Winqvist, lives close to a small town called Tranås, in the Swedish countryside. He studies at Chalmers University in Gothenburg, which is approximately 300km (186 miles) away from his home.

He found his tractor in a barn in January 2009 after the petrol Ferguson TE-A20 had been advertised in the local papers. Håkan thought it was an ideal project and would also complement his Russian off-road cars and vintage 1951 Opel Olympia.

The tractor had been standing in a barn for a very long time. He is only the third owner since 1950 and, surprisingly, the original papers still exist.

Good condition

Although the engine was not running, the rest of the tractor was in very good order. He had to remove the cylinder head and replace a valve and valve spring. He checked over the rest of the valve mechanism and re-ground the valve seats. The tractor is equipped with a Holley carburettor, which also needed repair.

All of the electrics were replaced, except the starter motor and the dynamo. The 6V control box was difficult to find in Sweden, so he had to order one from the UK. The round silencer that was fitted when he found the tractor is to be repaired, but in the meantime he uses a VW Caddy silencer that, conveniently, happened to be available.

In addition, Håkan had to replace some seals, bushes and the front axle pivot pin. Finally, he cleaned the tractor, stripped it to bare metal and re-sprayed it with two-pack paint. It is perhaps a bit more shiny now than it was when it was new, but he likes it that way – nice and clean!

Pleasure and work

The tractor is mostly used for relaxing trips in the summer, but also takes part in ploughing competitions and is used for snow ploughing in the winter.

In the small village where Håkan comes from there are actually three grey Fergies! An amazing total when you consider that there are fewer than 50 people living there. The youngest Fergie owner is Johan Fransson, who is just 10 years old. Not bad to have your own tractor at that age!

There is also a local club for old vehicles called "Ydre Veteranmotorklubb", of which he is a proud member. The club restored a Fergie as a project and it was offered as the first prize in a raffle. ■

The Swedish scenery is just stunning near the small town of Tranås.

As an avid photographer, Håkan is always keen to picture his tractor in the Swedish countryside.

The team line up – ready for the off!

The Flying Fergies

The Red Arrows display team practises over Suffolk: so do five Fergusons.

Two years ago members of a restorers club were trying to think of a new way to raise money at their club rally because a friend had been diagnosed with Motor Neurone Disease and they thought the charity could do with some extra funds.

After a lot of discussion and a fair amount of beer, they had a plan: there was nothing new in having a line-up of tractors but what if the vehicles did some sort of display like the Red Arrows?

The team was assembled, but how would it all work? How many tractors did it need to look good?

Five seemed to be the optimum number. Any less and it would not be exciting, any more and it could become large and uncontrollable, even dangerous. Also there just happened to be five drivers readily available, which helped make the decision. They were father and sons Keith, Gavin and Luke Argent and Dick Warren and Paul Lane – all Ferguson fanatics.

All they needed now was some sort of co-ordinator/commentator to think up routines and then wave his arms and shout when it all went wrong during practice. The task fell to the only team member who didn't have a Fergie, Jerv Jordan, who was born and bred a Ford man.

He was already well-known for his arm waving abilities and also for commentating at various shows and rallies.

The choice of tractor was fairly simple, as the Argent family worked daily at repairing and restoring Fergusons and had a few TE-20s available and the other two already owned TE-20s. So the team was built up of four TE-20s and a grey/gold FE-35.

Next it was necessary to work out some kind of routine. The Red Arrows display was studied but because of the inability of the grey Fergie to leave the ground by more than a couple of inches, aerial stunts were dismissed.

The White Helmets motorcycle display team was also observed but, unlike the Triumph motorbikes they ride, a Fergie will only reach around 10mph – and that's at full throttle with a following wind and the toolbox lid shut. However, the Flying Fergies did have one small advantage over the Red Arrows and the White Helmets, they could go forwards but they could also go backwards!

With this in mind the various elements of the display were planned – firstly on paper – which seemed safest.

The first few practices went surprisingly well, carried out at a slow crawl to help avoid any disasters. The tractors received no damage and the drivers showed signs of excitement and confidence. It was a very sharp learning curve, with some 18 different elements included in what amounted to a 30-minute display.

Practice sessions were broken down into a few elements at a time, then pieced together as ability and confidence grew. An interesting lesson was learnt here. Confidence and ability don't always go hand in hand. Lots of ability coupled with little confidence can be overcome by praise. However, a lot of confidence and little ability is a different thing altogether. It's all very well going like hell, but quite dodgy if you don't know where you are going or why.

The display was planned to include not only formations, which would show the team's ability at precision driving, but also daring stunts, performed both as a team and as individuals. Added to this was a fairly large helping of tomfoolery in order to create appeal across a wide range.

The more the team practised, the better it got, and the better the members got, the faster they went. This not only raised their confidence, but started the adrenalin flowing. The result of this was good to see. High-speed crossovers were performed with hair-raising precision, and excitingly individual stunts were developed.

Once the display itself had been sorted out, it was time to develop some sort of image. It was decided to mimic the World War I flying aces, so flying helmets and "sticky-outy" white silk scarves were made for the team. Now its members really looked the part, albeit in a slightly stupid sort of way.

Keith "Flash" Argent pulling out a trademark move.

Following many hours of practice, both in the field and at the pub, the time for their first display was fast approaching. This was to be at the Monks Eleigh Bygone Collectors Club annual rally in June 2007. The team were all members of the club and the crowd loved it.

The team entered the ring in their opening formation, the Diamond Arrival (since copied by The Red Arrows!) and, after one full circuit, continued to change positions skilfully to carry out five other formations including the Pitchfork, the Flail and the Windmill.

Throughout these manoeuvres Grey Gold leader Gavin "Spare Part" Argent, driving his FE-35, led the team. ➥

Grey/gold leader Gavin "Spare Part" Argent studies the routine beforehand.

Luke "Crash Test" Argent unveils his signature move to stunned onlookers.

Following this, the four other team members dispersed to their home corners to perform their individual stunts. These were Flying Officer Keith "Flash" Argent – bum on bonnet; Flying Officer Luke "Crash Test" Argent – wheel footy; Flying Officer Dick "Bunny" Warren – look mum no hands and Flying Officer Paul "Santa" Lane – breakdown.

It would be unfair to reveal the rest of the team's display; suffice to say it involved its police support team PC Plonker, a troublemaker – the dreaded Henry Von Dexta Major – and the lovely Nurse Gadget.

Perhaps the highlight was the moment when Grey/Gold leader decided he would like to do a jump on his FE-35. Members of the public proved reluctant to lie at the end of the jump ramps, however, so it was down to the team members.

Or perhaps it was the moment when they had completed the Four-Way Crossover and embarked on the Suicide Alley run.

Since that display the Flying Fergies have performed at other shows and introduced a Ring of Fire.

However, Jerv Jordan says the highlight of the Flying Fergies' career so far was when the team handed over almost £1,400 (the proceeds of its 2008 season) to representatives of The Motor Neurone Disease Association.

The team members' friend, who suffered from the disease and was their inspiration, died on 23 December 2007 leaving a wife and family, so the Motor Neurone Disease Association will always be the Flying Fergies' number one charity.

Jerv Jordan said: "The team is looking forward to the opportunity to raise even more money for this important cause. We are totally non-profit making and are always open to offers of sponsorship, however large or small. It is a costly business to move tractors and effects around, and fuel also does not get any cheaper. At the moment many of the costs are paid out of the team members' own pockets, which does place restrictions on what we can do.

"With more support we could, I am sure, raise a lot more money for this very worthwhile cause, while entertaining a lot of people with a unique tractor display."

* Jerv Jordan (Flying Fergies co-ordinator) can be contacted by telephone (01787 227695) or e-mail (jerv@jordanatveng.com) for booking details or donations and offers of sponsorship. ■

The team mid-crossover... the epitome of concentration.

David has established a well-deserved reputation as a top-notch professional restorer but he had fully intended to keep this TE-D20 for his own collection.

A surprise Ferguson

This TE-D20 is full of shocks, as Gary Connolly discovers.

In recent years the market for vintage tractors has scaled new heights and there have been many stories detailing the latest record price paid for one tractor or another.

Despite the credit crunch the right tractor is still attracting good prices at sales around the UK and Ireland. However, despite my faith in a good market for good tractors, I was more than a little surprised when I heard of the price paid for a Ferguson TE-D20 in Omagh at a public auction in 2009.

The Ferguson TE-20 is an ideal introduction to the vintage tractor movement. Cheap to buy, in plentiful supply and with an ever-growing range of replacement parts, a TE-20 is within the capabilities of most enthusiasts.

Visit any rally around the country and there's a fair chance that the TE-20 will be the most numerous tractor on display. It is, however, these factors that keep the value of such models relatively low and even restorations of a high standard struggle to achieve prices of in excess of £4,500.

All my preconceptions vanished however when I heard that the subject of this feature had made the princely sum of £8,639 when VAT and auction fees were taken into account.

To my knowledge, this is certainly a record price within the UK and Ireland and perhaps much further afield?

Some of my surprise abated when I heard that the tractor had been restored by David Graham, of Dromore, Co Tyrone. David will be familiar to readers of *Tractor & Machinery* and his expertise as a tractor restorer is well recognised in Northern Ireland and beyond.

I caught up with David and enquired after the mystery buyer who had paid a record price for his tractor. The well-pleased new owner turned out to be Stephen Todd from Coleraine but it transpired that there is more to this story than just a tractor fan happy with his new acquisition. ➥

Buyer and seller pleased; David (left) did the work while Stephen paid the record price.

It's fair to say that Stephen Todd does not fit the stereotype of a typical vintage enthusiast. A professional landscape designer and entrepreneur, he is a specialist in public art and is responsible for, among other exhibits, the landmark sculpture recently erected on the outskirts of Whiteabbey in Co Antrim. He's also the only tractor enthusiast that I know who drives a Porsche 911 Turbo! However, you don't have to scratch too far below the surface to discover his farming heritage and his true interest in tractors.

"I spent a lot of my childhood on a farm with my grandfather and an uncle down in Ardstraw and that's were the secret passion for tractors comes from. All my friends think I'm absolutely mad!"

Stephen's first tractor was an MF 135 and he freely admits that its initial function was stress relief. He took great pleasure in driving it around the paddock at his home, finding it a welcome relief from the trials and tribulations of his professional business.

He decided that he would like to have the tractor restored and, following an initial discussion with Brian Jenkinson of BJ Tractors, Preston, he was put in touch with David Graham.

Now for the twist in the story. While Stephen was keen to have the tractor restored, the entrepreneur in him rose to the surface and he got to thinking that there was an opportunity to meet a gap in the market, namely a novel video production of the restoration process.

When he saw the quality of David's work and his approach to restoration, Stephen knew he had found the right person to complete the video-based restoration.

Stephen has since completed the DVD, fronted by Emma Louise Johnston, and launched a useful website: www.reborntractor.net. David and Stephen are passionate about the concept and hope that they can widen the appeal of tractor restoration, while providing education and entertainment.

It was when discussing the tractor restoration video that Stephen first came across the restored TE-D20. David mentioned in passing that he had entered the tractor in the annual vintage auction organised by Omagh Auction Group. While Stephen made no comment, as such, he was very impressed by the quality of the tractor

The heat shield is one clue to the tractor designation – TE-D means that the tractor is designed to start on petrol but switch over to the less expensive tractor vaporising oil (tvo).

Gleaming paintwork in the autumn sunshine – it's not hard to see what attracted Stephen Todd to this tractor.

and he inwardly resolved to buy it.

For his part, David was a reluctant vendor – he saw the auction as a way to advertise his services without necessarily having to sell the tractor. He said: "I set a high reserve on the tractor and I fully expected to bring it home with me."

Once entered in the auction the tractor created quite a bit of interest and it was no surprise that the auctioneer decided that it should be the last lot of the day. Imagine David's surprise when the opening bid of £5,000 was higher than his reserve price!

True to his resolve, Stephen was the highest bidder. "It's not that I got carried away. I've a lot of experience buying art at auctions and I knew what I was doing."

Stephen knows that he has neither the time nor the technical expertise to restore a tractor to this standard so buying was his only option.

As he had donated his original 135 to the restoration video project, he decided that he'd like another one for home and saw what looked like a prime example advertised on eBay. The pictures looked good and so he made the winning bid but when the tractor arrived at his house he knew he'd made a mistake. The tyres weren't right and a cursory glance at the paintwork indicated that the re-spray was of poor quality.

He's philosophical about it all the same and hopes that his experience will be a useful lesson to others contemplating an online purchase. ■

A clever touch – the red lamp gives a rear projection while the side of the housing is cut away to illuminate the registration mark.

Buyer's guide: TE-D20

Andrew Hall advises on what to look for when buying a TE-D20.

The Ferguson TE-20 range of tractors played a significant part in shaping the future of farm mechanisation when introduced to the world back in 1946. The 'TE' designation stood for 'Tractor England' as a result of the manufacturing arrangement between Harry Ferguson and Sir John Black of the Standard Motor Company. This followed a previous agreement between Harry Ferguson and Henry Ford which had resulted in the production of the Ford 2N and 9N in the USA.

The first of the English models was the TE-20, which ran on petrol and used an American-built Continental engine. This was followed by the TE-A20, which also ran on petrol, but used the newly-designed Standard-built engine.

The TE-D20 followed this and also used the Standard engine, but was designed to run on tractor vaporising oil (tvo). In 1951 a diesel model, known as the TE-F20, was introduced to take advantage of the rise in popularity of diesel fuel. There were other designations for narrow, industrial and lamp oil-fuelled models.

The subject of this article is to be the TE-D20 and I hope to provide some information and guidance for potential buyers of this popular model.

The reason for producing tractors to run on tvo was due to previous concessions on petrol prices for agricultural use ending, which meant that straight petrol machines

Tractors in original condition, particularly those with lighting kits, tend to attract most attention at sales and auctions.

When running the engine, oil pressure should be between 40 and 60psi. Temperature should be maintained in the green sector of the temperature gauge, in order for the machine to run efficiently on tvo.

Left-hand view of the engine. From 1951, 12-volt electrics were introduced when the diesel TE-F20 arrived on the scene.

were no longer as economical to run. However, one disadvantage of tvo is that an engine cannot be started on the fuel and petrol is required to warm the engine prior to turning over to tvo.

1949 saw the first of the TE-D20s and they differed from the TE-A20s in that they had a two-compartment fuel tank with a two-way fuel tap, a fuel vaporiser (which combined the inlet and exhaust manifolds to pre-heat the incoming fuel/air mixture) and a lower compression ratio to accommodate the tvo. Also the cylinder bore diameters were increased from 80mm to 85mm. Tractors equipped for tvo are easily identified by the aluminium heat shield on the right-hand side of the engine.

Engines

The engine of the TE-D20 models was designed and built by the Standard Motor Company and was also used in the Standard Vanguard cars of the period and the Triumph TR2, 3 & 4 sports cars. Although the internal components are similar, the cylinder blocks were cast differently for the tractors and included an integral flywheel housing to allow for the unit construction of the tractors, unlike the car engines, which had a pressed steel plate. They are not interchangeable.

If looked after, the engines are fairly bulletproof and should give years of trouble-free service. One problem with tvo engines is that of engine oil dilution – partly due to turning over to tvo too soon and contaminating the oil. The service period for a TE-D20 is 60 hours – approximately half that of a petrol engine.

A good quality 20-50 detergent oil should be used these days. Originally, straight mineral oils were employed with differing summer and winter grades. If an engine is known to have been running on straight mineral oil it should be stripped down and cleaned internally before using a modern detergent oil, as this may pick up sludge deposits and block the oilways! Oil capacity for these engines is 12 pints.

When inspecting a tractor with a view to purchase it, the engine should run quietly with no knocking and should have oil pressure of between 40 and 60psi. TE-D20 tractors are fitted with engine temperature gauges and these should be at the operating temperature before turning over to tvo.

If correctly set, the governor should provide smooth running with no 'hunting' and when idling the air blast from the cooling fan should be heard clearly. The correct carburettor for the TE-D 20 is the Zenith 24T2 and has an oil bath air filter to protect from airborne matter.

The cooling system contains approximately 16 pints of coolant and should contain at least one-third antifreeze. This not only protects from frost but also from corrosion of the waterways.

If the tractor being viewed requires an engine rebuild it is well within the scope of somebody with reasonable mechanical knowledge. No special tools are necessary and there is no need to re-bore the engines as they all have replaceable wet cylinder liners.

The term 'wet' comes from the fact they are fitted directly into the cylinder block and are in direct contact with the coolant. They do have paired figure of eight seals at their bases to seal the coolant from the sump and these should be replaced when renewing the liners. Parts for these engines are plentiful in supply from many aftermarket stockists.

Right-hand view showing heat shield on manifold to cover the vaporiser for tvo operation. Note also the heat shield on the exhaust pipe. Many of these have long since disappeared!

Clutches

The clutches fitted to all the petrol or tvo models are the 9-inch diameter single plate dry type made by Borg & Beck. TE-20s were designed before the days of dual clutches and the clutch set-up is simple in design and operation.

This tractor has aftermarket footplates and leg shields on the mudguards making for safer operation!

The dashboard of all the TE-D20s had an oil pressure gauge, a coolant temperature gauge and the air intake through the dashboard.

Sometimes, on older hard-worked machines, oil may find its way onto the clutch linings and cause sticky operation of the clutch. This can come from either the rear oil seal of the engine or the front gearbox seal. There is a tendency for the clutch to judder sometimes when being engaged slowly. This does not mean there is oil on the linings but is due to the direct design of the clutch linkage.

Replacing the clutch requires the tractor to be split at the housing and supported on a splitting rail.

Gearboxes

All TE-D20 tractors have a four-forward and one-reverse speed gearbox. The gear lever also serves as the starter control to prevent starting while in gear. Gearboxes are fairly robust and have no particular faults, although well-used machines may have a sloppy gear lever action due to wear on the selectors. Selector forks have been known to break after long service but are readily available, but do require the tractor to be split between the gearbox and rear axle to fit them!

Gear ratios are quite high and some machines may be fitted with reduction boxes sandwiched between the gearbox and rear axle. Tractors with these are worth seeking out as they provide eight forward and two reverse speeds, as well as the ability to use a tractor-mounted rotary cultivator.

Lubrication is shared with the rear axle and hydraulic system and contains 5 gallons – originally a straight SAE50 oil, but these days Universal tractor oil will do the job. Be aware that this is of lighter viscosity and all seals must be in good order to keep the machines oil tight!

Rear axles

The rear axles are well engineered and present no particular problems. When inspecting a tractor it is worth looking at the rear of the brake housings for evidence of oil leaks, as this is not uncommon. Oil from the rear axle may enter the brakes and render them inoperative.

Changing the axle oil seals requires the bearings to be removed from the half-shafts and the use of a press. Some tractors have additional seals between the half-shafts and the axle casing and these are easier to replace.

Brakes

Girling 14-inch diameter (2-inch wide) drum brakes are fitted to all TE-20 tractors. Provided they are adjusted properly and do not suffer from oil contamination, they are adequate for their purpose. A single brake pedal operates the main brakes and left and right-hand pedals operate the independent brakes for slewing the tractors in tight spaces. Access to the brakes is easy after the rear wheels have been removed. A parking latch operates on the main brake pedal and is spring loaded.

14-inch drum brakes are fitted to all TE-20s.

The design of the front axle allows for track width adjustments from 48 inches to 76 inches.

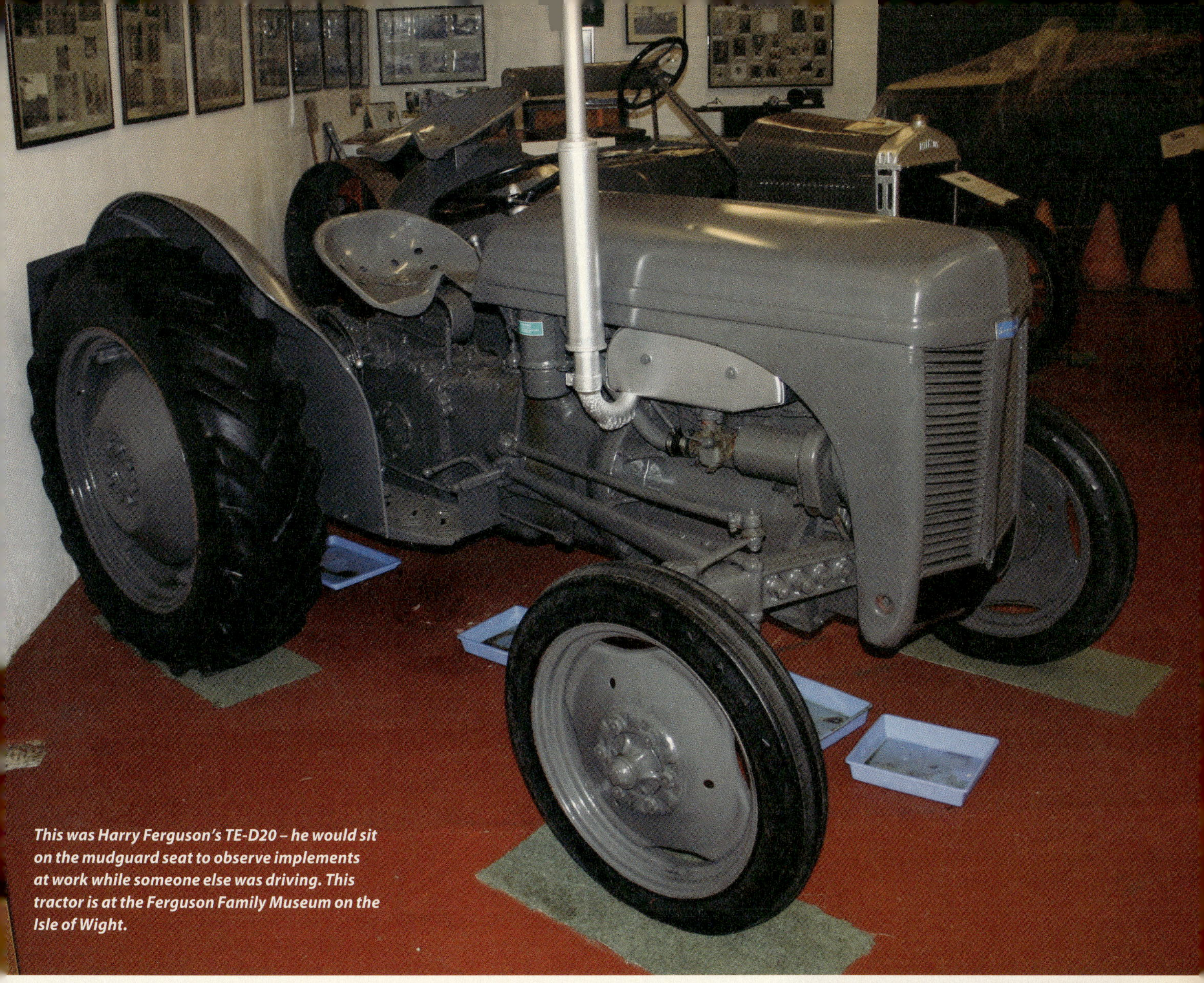

This was Harry Ferguson's TE-D20 – he would sit on the mudguard seat to observe implements at work while someone else was driving. This tractor is at the Ferguson Family Museum on the Isle of Wight.

Steering and front axle

The steering consists of a central steering box and two drag links operating each of the front wheels. The operation of the steering should be light and positive with no undue play present in the steering box.

Oil seals on the steering box drop arms are prone to leaking after years of service but are easily replaced. A filler/level plug is fitted on the centre of the steering box and should be filled with good quality gear oil.

The ball joints on the drag links should have minimal play and the rubber gaiters in good order. Regular greasing will keep the joints in good shape but replacement joints may not have grease nipples fitted and are sealed for life.

The front axle is constructed of three pieces and allows for track width adjustments from 48 inches to 76 inches without the need to adjust steering linkages, although the wheelbase of the tractor reduces slightly when on wide track widths. Regular lubrication of the pivot pin from an oil can through the starting handle hole will prolong its life.

Power take-off and hydraulics

All Ferguson TE-20 tractors have the drive to the pto shaft and hydraulic pump via the main traction clutch and this means there is no live clutch or hydraulics – confusing at first, but you soon get used to it.

To maintain the drive to pto-powered implements or to the hydraulic pump when stationary, the tractor gearbox should be put into neutral and the clutch re-engaged.

Live drive to these areas did not come until the arrival of the FE-35 tractor in 1956. Another point to note is that the pto shaft needs to be engaged to provide power to the hydraulics. This is a feature of all TE-20s and is a bit of a drawback, but is not confined to Fergusons as it is also a feature of Fordson Major E1As and early Nuffield Universals.

When using machines such as topping mowers, allowance should be made for lack of continuous drive by exercising a little anticipation when encountering heavy loads on the tractor.

Another point is that Ferguson TE-20s have a 1⅛"-diameter pto shaft that is smaller than the 1⅜" found as standard on other tractors.

This may be overcome by fitting an adaptor over the shaft at the cost of about £12-£15, or better still replacing the whole shaft with the larger size which costs around £80. This is the better option, as the adaptors are prone to breakage if overloaded!

The hydraulic system is what Ferguson tractors are all about and the TE-20 range consists of a category one-sized three-point linkage. Hydraulic systems are reliable provided the oil is kept clean, as there are no filters in the system. The system is ➡

Parts suppliers

Fergiland
Melbourne, Derbys.
Tel: 01332 862972
www.fergiland.co.uk

Forest & Farm
Brugh, East Yorkshire
Tel: 01430 448150

Ron Greet Tractors
Totnes, Devon.
Tel: 01803 812269
www.rongreet.co.uk

Carol Thulbourne t/a Carol (Spares)
Nr Downham Market, Norfolk
Tel: 01366 377464

Vintage Tractor Company
Newport Pagnol, Bucks
Tel: 01908 611864

View of power take-off shaft and pick-up hitch hook. This tractor has the larger 1⅜″ diameter pto conversion, which is more durable than the adaptor.

operated by a single lever and provides draft (depth) control to soil-engaging implements.

There was no position control provided on TE-20s for non-soil-engaging implements. If required to hold an implement at a specific height, it is necessary to use a support chain in conjunction with a toothed mounting plate on the top link. Two diagonal tappings in the base of the hydraulic pump pedestal, one on the front right-hand side and the other on the rear left-hand corner cater for external services.

Wheels and tyres

The standard wheel equipment for TE-20s is 4.00x19″ for front and 10x28″ for rear wheels. Rims are fairly durable, although many are showing signs of corrosion these days – bearing in mind the last tractors came off the production line in 1956. This is not a problem as there are plenty of second-hand rims available, as they are common to later models. New rims are also available from many spares stockists.

Spares availability

The popularity of these machines means that there are many sources of spares available and most parts required for a rebuild are fairly easy to obtain. In recent times reproduction bonnets have been made available and new Zenith 24T2 carburettors have been reproduced too. There are few excuses to display an incomplete TE-D20 tractor!

Conclusion

Whether it's for work or play a TE-D20 is a useful and economical tractor to operate, provided the operator is prepared to allow for power limitations. However, a wide range of power-matched implements were made for these tractors and it can be quite fun seeking them out to add to your collection, making the Ferguson tractor extremely versatile.

Prices vary considerably and range from £200 for a tractor needing considerable work to over £2,000 for a fully-restored example. Due to their simplicity, lack of dual clutch and position control, they rarely fetch the prices of the later 35 models, but are still fun to own! ■

The alternative to the nine-hole drawbar is the pick-up hitch assembly. Note also the sliding top link, which was standard fitment to TE-D20s. Some of these top links have a toothed rack for chains to fit in order to support mowers etc.

Recent prices at farm dispersal sales and auctions

10 May 2008; Cheffins sale, Northampton: £550.

7 June 2008; Cheffins sale, Barleylands, Essex: £1,000.

21 June 2008; H&H sale, Appleby, Cumbria: £850.

31 January 2009; HJ Pugh sale, Somerset: £400.

Alec Fairbrother on the TE-L20.

A birthday present

Peter Love visits Alec Fairbrother in Leicestershire, a keen Austin 7 owner, who returned to his roots when he purchased a lovely June-registered 1952 Ferguson TE-L20 as a birthday present for his wife Maureen.

As a boy, Alec learned to drive on a Fordson Standard N and joined the Ferguson clan later. After leaving the land and spending time at the Eastwell Iron Company quarry complex, his next career move involved working with boilers and lasted for 27 years.

Alec is also an Austin 7 enthusiast, owning three examples dating from 1931, 1933 and a 1936 two seater. Maureen, his wife of more than 40 years, is a regular assistant with his mechanical endeavours.

Maureen's birthday is on 17 December and Alec is always keen to provide her with something special as he does not want her to miss out, as do many of those with birthdays close to Christmas. In 2008 he asked her what she wanted for a present and when she replied: "something small and different" little did she know what that was taken to mean!

Just the ticket

Alec was with the 'Tuesday Club', members of which are also all members of the Vale of Belvoir Vintage Tractor Club. Alec Lord, a Ferguson man through and through, and Colin Cooper are the principals among a group of retired people who keep active by working on projects together and have often encouraged Alec Fairbrother, as a tractor owner, to join them in the regular mix of camaraderie and banter. While on a lunch break looking through the *Tractor & Machinery* classifieds, the Tuesday club gang found just the thing for him, and possibly Maureen.

Responding to an advert meant a trip to Wolcott, Lincolnshire to look at an older restoration, a May 1952 Ferguson TE-L20 vineyard model and the fourth off the production line.

The engine of this 37-inch track machine ran, but transmission problems meant it stuck in gear at times. The team reckoned it could fix the problem, a deal was agreed, and in November 2008 the tractor moved counties. ➡

The front end on the vineyard model. This is a different arrangement to the regular narrow version.

Compact

The TE-L20 is certainly a low-line machine. It's more than three inches closer to the ground than the standard model and has 24-inch wheels in preference to the standard 28-inch versions. The bonnet is four inches longer than a standard example. The vineyard model is a useful option to consider if space and ease of handling are important issues, although it is not a very good option for ploughing. In a Ferguson line-up, it's the vineyard model that always attracts attention.

Around September 1948 Ferguson introduced the 42-inch narrow tractor. The modifications were carried out by the long-gone, Kent-based Lenfield Engineering. At the Sittingbourne branch they converted the tractors on behalf of Harry Ferguson Ltd. These included fitting shorter trumpet housings, which were believed to be supplied by Wellwinch Engineering. Obviously new half shafts were required, but these may have been supplied by Banner Lane.

The brake mechanism was modified, along with the linkage and various other components. The tractors even featured steel hub caps, which inevitably got bashed about in routine use. These were coded as the TE-C20 model with the Standard engine. A few were built with the Continental engine and this model was listed as the TE-B20.

Still too wide

For the berry farmers of the Clyde Valley in Scotland these tractors were still too wide. As a result, a number of influential people tried to persuade Harry Ferguson to make an even narrower version, but to no avail. John Reekie at Lochland, Abroath came to the rescue and produced just what was wanted from his family's Ferguson dealership, Farm Mechanisation, a business he had established with his brother in 1947.

His fruit tractor modifications included cutting and re-welding the half-shafts, a process that met with Harry Ferguson's disapproval, but the tractors came on the market in June 1949 and were a great success.

These tractors are now highly sought-after, especially as no two are quite the same. Some are believed to be as narrow as 32 inches, but adjustable to 56 inches. There is also some interesting correspondence when Reekie requested approval for promoting and selling his 'berry' conversions, as they were to become known in Scotland, and was rebuffed by the Ferguson company on many occasions. Reekie also produced suitable Ferguson-style narrow implements to go with the tractors, including trailers.

Tailored implements

It is suggested that as many as 600 of the Reekie tractors were built, but with the introduction of Ferguson's own TE-L20 vineyard model (with the Standard vaporising oil engine in June 1952) people lost interest in the conversions. However, Banner Lane also offered the vineyard in petrol (TE-K20) and zero octane (TE-M20) versions.

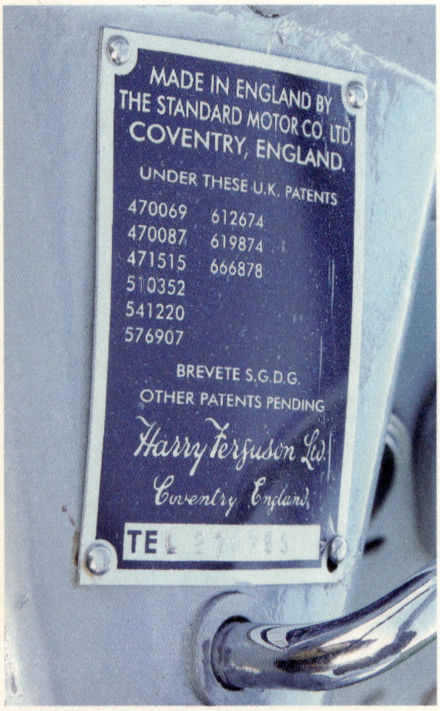

The identification plate on all TE-20 tractors is just above the throttle lever on the steering shaft bracket.

The finish on the tractor is stunning, with even those hard-to-reach places looking good!

Alec's TE-L20, from the first day of production, came home in early November and was tested around the fields in Leicestershire with the 'gang' in attendance. Maureen came down to see the tractor, decided to have a drive, instantly fell in love with it and decided to call it 'June' as that is when it was registered. Alec thought this was the perfect gift for her forthcoming birthday, as she had asked for 'something small and different' and the tractor fitted the bill.

Project TE-L20 begins

That same day the tractor was stripped and within a few weeks the gearbox had been replaced by a good second-hand unit. The clutch went straight back in as it, and the flywheel, were in excellent condition.

A number of seals were renewed, including that for the pto at the rear end. While the engine was being looked at, all four wheels received new tyres and tubes. After cleaning and pressure testing the radiator, the cooling system received a new water pump. The steering needed some work with new thrust washers on the kingpins and a new track rod end and some protective boots.

The engine appeared to require little attention, but the ignition eventually did give problems. Inevitably, there were other minor details needing attention. Alec also had the hand throttle re-chromed along with the gear lever knob.

The tractor had been shot-blasted and the team applied three coats of zinc-rich acrylic two-pack primer. The mudguards were repaired and welded where necessary. ➡

The wheel centre and rims also needed full restoration. The TE-L20 carries 9.5-24 tyres, which help to lower the tractor.

The owner has arrived! Maureen enjoys 'June' the TE-L20.

After some filling and a lot of rubbing down the skid unit and panels all came together for the grey top coat to be applied, wet on wet. A new wiring loom was fitted and the original Jack Olding dealer plate went back on as well.

Making its debut

It was intended to debut the tractor at the 2009 Belvoir Castle event in May. On the Thursday before the show it was ready to start up but would not fire! Panic set in, but after replacing a condenser and some leads – away she went! A later drama, while in the parade ring, led to the replacement of more ignition components.

The tractor is a credit to the efforts of the team, particularly Colin Cooper and Alec Lord. Maureen thinks it's lovely and she's right, emphasised by the fact that the TE-L20 attracted admiring looks at the Newark Vintage Tractor Show in November 2009, on the Vale of Belvoir stand. What follows June? July, of course, but that's a red tractor from another make! ■

The Tuesday 'gang' working on another project.

The rear linkage was completely overhauled.

Bob McAllister and his award-winning Ferguson TE-F20.

A Ferguson tale

Bob McAllister's TE-F20 is one of the best of its type on the Northern Ireland rally scene. Gary Connolly listened to Bob's story of the restoration of a family heirloom.

"On 22 August 1952, John Brook and family bought a TE-F20 diesel tractor from George Henderson Ltd, Fingston, North Berwick. The salesman was George Milton of Kingston who is, as far as I know, still alive and well. At that time the tractor would have cost about £550.

The Brook family had a large farm of 550 acres and owned around six other tractors – including Fergusons and Nuffields. The new tractor was given plenty of work including ploughing, potato harvesting and silage harvesting.

The family kept the tractor for nine years before Andrew Brook sold it around the time of his father's death. The tractor was imported to Ireland and was offered for sale by Sam Hazlett Tractors in Moneydig, Garvagh.

In 1961 my father, Gary McAllister, bought the tractor from Ian Wilson who was in partnership with Sam Hazlett. The tractor cost £210 – a lot of money back then – and I still have the original receipt given to my father by Ian.

The tractor did a great deal of work on our farm – ploughing the fields, bringing in ➥

Attention to detail is what sets this restoration apart from many others.

the hay and digging potatoes. It was also used in the moss (the local name for a peat bog) to bring home the turf for burning over winter. The tractor was very good in the moss because it was light and did not sink when pulling a sledge loaded with turf.

At that time, a lot of our neighbours didn't have a tractor and our mother and father loaned our tractor to other farmers in the area. That was the way it was in those days – one person helped another. Unfortunately, my father fell ill and we were too young to work the land, so my mother and father sold the farm in 1970. Luckily, they kept the tractor to work in the moss.

Sadly on 3 March 1990 my father died, aged 66, leaving me the family home and the tractor along with it. The Ferguson had been kept for years in one of the sheds, as it was no longer needed.

In 1997 I went to a vintage rally in Limavady, just to see some old tractors and have a day out. As I was walking around the show, I spied a small grey tractor, which turned out to be a TE-F20 – just like the one I had in the corner of my shed. The tractor looked very, very nice and was well restored. It really caught my eye and at that point I was hooked!

I had lots of visions going around in my head as I stood and looked at the tractor. I said to myself, "my dream is to some day have my tractor look like that, no matter what the cost." I think it's because my tractor holds a lot of memories for me of my father – he and I were very close.

So, I set to work on the Ferguson. The first thing I did was take a photograph of the tractor. The following day I took the tractor to work and pressure-washed it three or four times to get it really clean. I then put it in the shed and stripped it completely – including the engine.

It needed a lot of work and I soon discovered that the engine would be hard to work on, although this didn't worry me. I am not a mechanic, but I am mechanically minded and that is half of the battle. The engine needed a complete rebuild, including new rings and bearings.

The next job was the bodywork. I cleaned it down with a wire brush attached to a drill. This work seemed to last forever but I think if it's worth doing, it's worth doing right. I fixed all of the oil leaks before thinking about spraying the tractor.

When I was ready to start spraying, I rested two ladders on empty barrels and hung every part I could get off the tractor from the rungs of the ladders. It took me over three days just to reassemble the tractor again.

I really did go to extremes when restoring the tractor – to put it back to the way it would have originally come from the dealer including, for example, restoring the original batteries and fitting a set of old original 10x28 Goodyear tyres.

The lighting set is quite rare – the single large light and the two sidelights were made by Lucas and were fitted to some of the early Fergusons. This set was fitted to my tractor by the dealer over 50 years ago.

I worked on the tractor night and day for over three years, sometimes until one or two in the morning, but I eventually got it finished, having done all of the restoration work myself.

By this time, my wife thought that I had fallen in love with the tractor and I suppose I could not blame her. I really enjoyed working on the tractor – I think because it holds so many memories.

The original fetaures, such as safety chain and battery covers, completed the job.

Achieving a good finish on the bonnet is imperative, as it's the first thing that people see. Look at the reflection of the headlight!

Bob likes to look after his TE-F20 and has even customised a twin-axle trailer to transport the tractor to events.

The very first show I went to was back in Limavady. One or two people that day thought the tractor was a 1953 model and I was not best pleased. When I got home I started to look for the tax book but, after looking the house up and down, I could not find it. However, in the process of searching, I came across some good news and some bad news.

I did find one of the old insurance policies relating to the tractor (dated 1973) from a company called Wallace Kerr of Kilrea, priced at £2.50!

The bad news, however, was that the receipt that I found from Sam Hazlett Tractors stated '1953 Ferguson tractor'. I was even more confused – what year was my tractor built?

I rang a company in England and gave them the serial number from the manufacturer's plate and, thankfully, they confirmed that the tractor was made in 1952. Had Sam Hazlett made a mistake? In those days, the year of the tractor wasn't that important.

I knew that the only way to solve the mystery was to get the tax book. I rang the Vehicle Licensing office in Coleraine to check if they could get me a tax book, but unfortunately they told me that their records only went back 30 years.

My mother told me that my father had got a tax book at the time he had bought the tractor, but it had got lost. Sam Hazlett Tractors had closed down in 1999 and the premises were now occupied by Torrens McFetridge.

As a last resort, I decided to ring the new owner and tell him of my search. Torrens told me that there was an old filing cabinet in the back and that he would have a look and give me a call. After 4 or 5 days, Torrens phoned me and asked me to call in and see him.

My heart started to beat a little faster as I went to see him – could this be good news? When I got there, Torrens produced a tax book and asked if it was mine.

I could not believe it – there on the table was my father's original tax book. My eyes lit up – it was like winning the Lotto! The date read 22 August 1952 and so I now take a copy of the tax book with me to all of the shows to prove that it is a '52 tractor and not a '53.

I have spent thousands of pounds on the tractor but, as far as I am concerned, it is worth every penny. I have won first prize at a number of events and rallies and these help make it all worthwhile. A valuable tractor – yes! Sell it? – Never!" ■

Bob admits that he has spent thousands restoring his father's tractor, but the expense was certainly worth it. It's a tractor to be proud of.

Pitch and putt Fergie

While on a short tractor holiday in Ireland, Steve Wright found a Ferguson TE-F20 still earning its keep, working every week to maintain a good pitch and putt course in Blessington, County Wicklow, Ireland.

We found ourselves with a bit of spare time on the day before the popular charity event, the Hillbilly Tractor Run. As we drove by the golf course I asked my son Archie if he wanted to try it out and he was very keen. But to our surprise just as we were nearing the finish of the 18-hole course we spotted, tucked away behind a tree, the giveaway colour and shape of a grey Fergie. It pays to keep a look out when on your travels, as you never know what you may find!

While it could probably benefit from a thorough service, this tractor has been a real reliable workhorse over the years for its owner, Paddy Keogh, who commented: "It just keeps doing its job, I keep promising to do something with it, one year maybe I will."

An opportunity

Paddy always fancied owning a grey Fergie, having seen so many of them still working the fields when he was a boy. The opportunity came to buy this one in 1987 from a local trader for IR£250. The serial number is TEF 311275, which means that the tractor was built in January 1953.

Not much is known about the first 34 years of the tractor's life, but over the past twenty-

Paddy Keogh puts the Ferguson TE-F20 through its paces on the pitch and putt course.

two years it has certainly continued to earn its keep. It's had a new battery and new wiring, but not a lot more. It weathers the elements outside all year round and is used every week to cut the greens with a ground-driven Ransomes Ipswich sports cutter – this is normally a three or four-hour job.

Ideal choice

The tractor is, of course, quite economical with fuel but Paddy does have to top up the oil every time he uses it, he gets through a litre a week. The radiator also has a slow leak, so this also requires filling up every time. The alternative to this little Fergie, says Paddy, could be a €30-40,000 modern tractor with a fancy cab and air conditioning. But he's happy with his Ferguson TE-F20, and we like to think that this is not just because it's cheaper! ■

Above: The Ransomes ground-driven cylinder mower is old, but still very capable.

Below: The Ferguson TE-F20 works hard and requires oil and water every time it is used, but it is still a viable alternative to a high-priced modern tractor.

The blade on the earth mover can be offset, angled up to 23 degrees left or right and the pitch of the blade can also be varied depending on the soil conditions encountered.

Moving the earth

Kim Jackson visits a collector with a passion for Ferguson tractors.

The centrepieces of Peter Harris's tractor collection are his grey and gold Ferguson 35s. But the first tractor he restored was a little grey Fergie – as was the last, which he took along to Tractor & Machinery's Tractor World show in Malvern.

Peter hails from Ellerdine, a few miles away from Shrewsbury. His family used to have a dairy herd of Jerseys but these have recently been sold and now Peter's father concentrates on rearing young stock. In this very rural and agricultural setting it was not very surprising that Peter grew up with a huge liking for tractors and anything associated with them.

His father's preference for tractors has always been Massey Ferguson, but Peter's specific interest pre-dates anything that he can remember on the farm – tractors which solely bear Harry Ferguson's name.

The latest project to arrive at Ellerdine was a TE-F20 and a Ferguson earth mover or, as most would say, a bulldozer blade. Even in the 1950s it would have been dwarfed on a building site, but Ferguson's sales team no doubt thought it would be suitable to help clear small areas on the farm, perhaps for a new shed. Since new sheds are not yearly occurrences on farms, the maximum action most would have seen was keeping a road clear during a snowy spell.

It was far from a top seller but 50 to 60 years on this makes it all the more desirable and, with most having seen little use, they are in good condition.

Peter did not find his at any of the local auctions he enjoys attending, but through information supplied by the mobile library driver. Peter went to Newport, Shropshire, and found the blade which had been purchased new for snow clearance. A surprisingly easy deal was struck.

When the blade was delivered, Peter was mainly pleased with the condition it was in; rust was showing but it was certainly not rotten. It was all complete and even had its original badge on. Peter's only real work on it was to smooth it down, before painting it.

In the past he would have used an ordinary wire brush, which would have taken a considerable time. This was a necessity because Peter can only do certain things with his hands and the co-ordination necessary to use power tools safely was not possible. A recent achievement, though, is that he has mastered a cordless drill which has a wire brush attached to it and, at a stroke; he has cut out a considerable amount of repetitive work.

Although there was a suitable grey Fergie to fit the blade on, Peter was looking for another project and bought a TE-F20 in February 2007 then spent the next eight months restoring it.

In his words "it was in pretty good condition, the engine was perfect." No spanners were needed underneath the bonnet; the main problem was with the back end. At some time in the past a TE-D20 axle had been fitted on to the back of the TE-F20, which meant the original battery tray could not be connected to the correct position on the axle.

Considerable thought was given as to whether the battery problem could be resolved with the current axles but in the end it was decided that to make a decent job for Malvern, the TE-D20 axle had to come off and a replacement TE-F20 axle found.

After the TE-F20 arrived in Peter's shed/workshop he decided that the first thing that needed doing was the front end and completely stripped it down. During the reassembly he opted for new bearings, bushes, pins and bolts. For Peter to do this, they must have been well-worn, as he shows a clear preference to use original materials. New 'shiny parts' are only bought if considered absolutely necessary.

After the main part of the tractor was rebuilt to Peter's satisfaction he then turned his attention to the back end, with considerable resourcefulness.

His disability means that he spends a large amount of time in a wheelchair. This is, no doubt, a major disadvantage most of the time but occasionally it does work to his benefit. Peter doesn't possess a crane but by using a trolley jack and suitable packing he turned his little grey Fergie round from his wheelchair.

The guts were then stripped and, again, more new parts were needed. The trumpet housing was replaced with a second-hand one along with new half-shafts and housings.

At first, finding suitable back wheels was a little problematic, as the TE-F20 had arrived at Ellerdine with odd wheels, one of which would originally have been supplied to fit on a Fordson Dexta. Peter advertised for one, but got no response and then remembered he already had one!

Many other readers have no doubt had similar experiences; when attempting a tidy out, they uncover parts and tools that went missing so long ago they forgot they ever had them. Goodyear tyres have since been fitted.

Peter manages all of the painting himself, from the bonnet through to the mudguards, and all of the skid unit in between. He has his own way of doing this and expertly holds and manoeuvres the paintbrush in his mouth.

It is certainly an education to go into his workshop and see this in action. But it ➥

Peter Harris's Ferguson TE-F20 was purchased with a sound engine, so no hammers and spanners work was needed on the engine mechanicals.

Everybody should go through life learning new tricks and skills. Peter's latest accomplishment is to master a rechargeable drill with a wire brush attachment.

At some time point the tractor was involved in a serious accident, resulting in a TE-D20 back axle being fitted. Peter is putting the undercoat on one of the trumpet housings.

The Ferguson earth mover is attached to the TE-F20 by a sub-frame to distribute the forces from the bulldozer back to the rear axles. The Ferguson brass plate gives the machine number as 1309 and model 528. A second plate below indicates that the earth mover was made under licence from Bomford and Evershed.

needs considerable patience; just to put one coat on the bonnet took him a full day.

In all, he goes round the tractor three times with red oxide, undercoat and top coat. Before the top coat is put on everything is rubbed down with an emery cloth, another job which to many is now just too laborious without the aid of power tools.

Full attention was given to lights and Peter priced up a new lighting kit at £400 and decided against one. Instead he sourced the parts needed from a mixture of places.

The tractor had arrived with just the brackets bolted to the side of the bonnet. There were no arms and no headlight in the middle, but there was a bracket for one. Pondering the lights issue, Peter took a photo of an original tractor and decided to make the arms himself.

To do this he folded metal around a piece of wood. Individual lights were then bought to go on the ends. Finally the much sought-after middle light was sourced from an autojumble.

Peter doesn't just work in metal; he is very pleased with a wooden battery cover he made himself. Again this is an item that is quite easy to source, but Peter thought the cost did not justify it. And as many a good craftsman can testify, it is much more satisfying to make something than buy it.

In all, the tractor took about eight months of work and the blade one month. The tractor with its earth moving attachment was then premiered at Malvern in March. Pete has already bought his next project, a narrow wartime Ford-Ferguson. ■

Peter is able to draw on the support of two good friends, Neil and Keith (the driver) for the more difficult jobs and driving the tractors at shows.

The non-standard seat is marginally more comfortable than the original.

55 not out in Kent!

There are many early grey Fergies out there, although many are in the hands of collectors or enthusiasts. There are many still at work on smallholdings or smaller farms. Paul Smith visits one TE-F20 that is still earning its keep every year at a small cricket club in Kent.

Throughout the 1950s, '60s and even the '70s, the fields and lanes of rural Britain echoed to the characteristic thrum of what many would argue was probably the most recognisable and well known of the Ferguson products, the TE-20 range.

Often the primary choice of workhorse for farms big and small, nurseries and even airfields, it wasn't only in Britain that this versatile and reliable tractor could be seen earning a crust in daily use. In its heyday, the tractor, in its many variations, was exported the world over with over half a million built.

Nowadays we are more likely to see the TE-20 variants restored to their former glory, gracing rally fields and show rings with many examples being nursed in respectable and well-earned retirement by the die-hard enthusiasts who keep the memory alive.

Breaking the silence

In a quiet corner of North-West Kent however, the wonderful throaty rumble of one of these machines may still be heard, being given an outing in regular and humble use each year throughout the summer months.

Bexley Cricket Club and Burnt Ash Hockey Club share a five-and-a-half acre field which sees a Ferguson pulling a three-gang cylindrical grass cutter during the season – helping to keep the grounds in first-class condition.

This particular machine isn't one that has been rescued, nursed and renovated either! This Fergie is a 1954-built TE-F20 type with the 4-cylinder Standard diesel power unit, which came to Bexley about 20 years ago and has been the principal power source for the weekly summer mowing programme ever since.

The tractor is almost as much a part of life at Bexley as the shouts of 'owzat' and post-match tea in the pavilion and its continuing employment, albeit limited to around ➥

Cutting in front of the pavillion.

seven months of the year, bears testament to the type's rugged reliability.

It may appear a little work-worn and somewhat down at heel but as Brian, the former head groundsman, explains: "She just keeps going and, although her work is limited to only one or two outings a week, depending on the weather, she serves us well and has never needed much attention." Brian even said that a puncture repair about two years ago revealed an original inner tube in one of the rear wheels which did need replacing, but the repairman was amazed it had survived that long!

Almost original

The 55-year-old is in almost original condition, the only concession to modernity being the replacement of the original metal seat with a newer plastic one which is apparently only marginally more comfortable. A new battery was needed for the current season but, apart from some annoying, though manageable, oil loss, very little maintenance is required.

On seeing the machine at work however, aficionados and sharp-eyed readers may well notice that this particular TE-F20 does not come in the characteristic, as-delivered, grey paint scheme, which constantly made the type instantly recognisable.

This Fergie apparently acquired its coat of blue a few winters ago when a young ground hand, finding himself short of tasks during a few days of particularly heavy snow, decided to fill his time with a spraygun! The chosen shade of blue does, however, suit the tractor quite well and gives it a unique and somewhat individualistic identity, and the ensuing years have resulted in areas where the original grey paint is again starting to show through again.

Each spring in Bexley the TE-F20 is brought out to begin the season's grass-cutting and it takes to the task extremely well. The model, of course, employs the original starting routine that uses the gear lever to fire the engine into life, complete with the distinctive burst of grey smoke from the exhaust.

The aforementioned oil loss apart, there are some quirks and signs of age, such as an almost complete lack of brakes (which apparently isn't a problem when the tractor is only used for cutting grass on a level field!) and a reluctance to engage reverse gear – solved by simply choosing neutral, getting off and pushing it backwards...

Although the tyres are more suited to industrial applications, they are acceptable for grass work.

Working together in perfect harmony – a good set-up.

A regular routine

Seeing this old Ferguson at work and hearing that wonderful gentle pulse echo across the neighbourhood as it goes about its weekly routine, it takes one back to those heady days when life was perhaps a little less complicated. There is the impression that those who operate and look after her are in fact keeping a little piece of great British engineering alive and well in workaday, un-restored condition and employing it for the purpose for which it was built, which somehow all adds to the appeal.

Will the Ferguson ever be retired? No doubt in time there will be a need to replace it with something more up-to-date, but in the meantime, oil leak and recalcitrant gearbox permitting, it may be that this particular tractor will be around and keeping the grounds in Bexley in top condition for a few more years to come. ■

Bruised and beaten, this small Ferguson still earns its keep in Kent.

Gary Houghton and the award-winning TE-F20.

Out of the ashes

Maurice and Gary Houghton are always on the look out for a potential project and the latest one came in the form of a 1955 Ferguson TE-F20 and a two-furrow general-purpose plough.

Maurice Houghton's employers were originally F A Standen, where he served his apprenticeship, and he stayed with the company when it became Thurlow Nunn Standen, retiring in 2001. His duties involved repairing Massey Fergusons all over East Anglia. Knowing of his interest in older machinery, in 1985 he received a tip-off regarding a 1955 Ferguson TE-F20 at Potton in Bedfordshire that had been "mildly burnt out".

The wiring loom had got hot and melted a plastic diesel pipe and started the fire. The damage was quite extensive, even including the steering wheel, burned inside the cab. All the radiator hoses were burned and the soldered joints melted on the radiator and fuel tank.

The panels were slightly damaged (warped through the heat) but Maurice thought the tractor (serial number 438477) was worth saving. He brought it to Huntingdon on 6 April 1985, without the cab, and parked it in the shed for future restoration.

Pushed into the workshop

The year 2005 marked the 20th anniversary of the tractor coming into the Houghton family's ownership and it was also the year when restoration work started. The tractor was pushed into the workshop and Gary started on the strip down of the engine (serial number SA68228E).

He marked all the head bolts so they could be returned to the original locations, but removing the cylinder head revealed a badly worn engine and the need for a complete strip down. Having had considerable practice, Maurice cajoled the pistons and the con-rods out of the bores, a job which has a reputation for not being one of the easiest, even if the crankshaft is out of the way. The rebuild required new liners and a light skim of the crank before all the bearings were replaced.

This is how it looked when the boys got it home after the fire in the early 1980s.

Everything covered

With new pistons and seals fitted, the bottom end was built up with a new timing chain. The oil pump was cleaned and thoroughly tested before refitting. The head needed a good de-coke and the valve seats were re-cut before new exhaust valves were ground in. While this work was taking place, the CAV BPE injector pump and injectors were sent away for overhaul. By the following month these were ready to be fitted back. Maurice advises caution when fitting this type of pump as the correct timing is critical.

Turning to the cooling system, Gary washed out the radiator header tank and managed to re-solder it back together. Pressure-testing it proved the job was a success. A new water pump went on to the engine along with new hoses and clips, but a new fan assembly had to be obtained as the melt-down had ruined the original.

For the fuel system, new pipes were made and again Gary got the solder gun out and sorted all the melted joints on the fuel tank before it was resealed inside. The Ki-gas tank had to be renewed and an original Ki-gas unit (pre-heat) was obtained, to be wired up later.

On and off

This was an on-and-off job over four years but the steering was tackled next. The steering box was in quite a state, something that is not common on these tractors, and the bearings were replaced, including that on the steering shaft, which took some time to do.

The centre pin in the front axle was replaced and re-bushed. The wheel bearings and king pins were sorted and re-shimmed and new boots went on the steering joints where required. This completed the overhaul, which transformed this area of the tractor.

However, the rear end needed more work than first thought and this started with new pins in the trumpet housings for the link arms. At the same time it was discovered that in the differential centre housing four gear teeth were broken, which was a surprise, and a firm conclusion on how this happened has never been reached. Luckily, a TE-F20 'scraper' was found locally to provide for repairs.

The clutch was fine so this went back in, the gearbox was in excellent order and the brakes were overhauled and refitted, with the cross shafts freed off, cleaned and re-assembled.

By this time the engine had been painted in rust protective paint so it was reunited with the transmission. The rear housing was wire-wooled down and sanded before the same paint was applied.

The starter motor and dynamo had been sent away for repair and came back useable, which was quite remarkable in light of the fire.

A new wiring loom and control box were fitted (and the correct type of original cable covering was used) along with the new gauges that had to be used with the new steering wheel. A hidden 'kill switch' was placed on to save anyone trying to start the tractor at an inappropriate time.

Turning to the batteries, new trays were fitted and a dummy 6V battery went on ➡

Progress! The tractor is seen in March 2009 with the engine sand-blasted and assembled and the back end connected again. Red oxide primer was applied before...

... primer filler went on prior to the skid unit receiving a top coat.

F A Standen was an extensive John Deere agent, during World War Two, but took to the grey Ferguson with vigour, as seen here.

one side and a powerful 12V the other with, of course, the authentic Bakelite-type covers.

Testing begins

It was time to test the tractor before the painting and final parts were added and this was done in the garden using a Ferguson seven-tine cultivator behind. Nothing untoward showed up during the process and that was most satisfying.

Running up a tractor in this way is essential before the final stages of painting are undertaken. On some occasions an oil leak, or some other problem, comes up that needs attention.

Back in the workshop various parts were taken off again as the painting process started in earnest, including the wheel assemblies. A pair of second-hand wheel rims with the original centres and a pair of second-hand Goodyear tyres with new tubes went on the back. On the front went the 4.00x19 diamond-pattern Goodyear tyres with new tubes on the original rims.

Regarding the panels, the bonnet was salvaged but new mudguard skins went on using the correct sized domed bolts and nuts, again something that is very important if you want your restoration to be right. All the panels and sundries were sand-blasted and rust inhibitor paint was applied followed by heavy-build primer.

After about seven coats it was time for a heavy rub down, that eventually led to 'Ferguson Vintage Grey' being applied in outdoor conditions. This paint dries quickly and came up to a standard similar to that in which the tractor would have left the factory.

Finishing touches

It was time to fit the lights which came from various locations, with the bonnet headlight coming from an autojumble. Last on was the rear plough light and its switch.

The new white exhaust silencer went on and to finish the tractor off, the correct size registration plate and bracket were fitted to

A new Ki-gas tank had to be made and the fuel tank was re-soldered.

Maurice had the F A Standen decals made.

The centre-mounted 'Cyclops' headlight.

Above: With the plough attached, the TE-F20 really looks the part. Gary and Maurice should be proud of their achievements.

OEW 274. A new manufacturer's serial plate, to replace the one melted in the fire, was obtained and fitted. The decals went on to finish off the tractor and it was ready for its first rally in late August.

The plough

At the same time Gary had picked up a Massey-Harris-Ferguson two-furrow general purpose plough (number 045802). It was a good £50-worth of spending, the shares were in good condition and the mouldboards were good. They were all cleaned up and stripped down and the slides and cross shaft were freed off and the skimmers got similar treatment.

A new set of the correct sized bolts and nuts went on before sand-blasting, followed by numerous coats of paint, well rubbed back. The result is excellent with its new commission plate and decal.

The father-and-son team's objective was not to produce a concours tractor, but one that was practical to rally and use. I feel that objective was achieved and the judge at the Haddenham Steam Rally, where the tractor finally debuted on 13 September must have agreed. It came away with the 'Best New Restoration' award.

It might have taken 24 years to restore the tractor, but it had certainly been worth the effort. ■

Many parts were re-chromed.

The hydraulics required little attention.

New ball joints went on to improve the steering.

A correct Ferguson seat!

They don't come this good every day!

A gift from Fred

Peter Love visited Neville Porter in Castlederg, County Tyrone, Northern Ireland to look at the lovely 1955 Ferguson that Neville has restored. Although at first he wasn't a true Ferguson fan, he undertook this mammoth challenge as an arrangement to the previous owner.

Neville Porter is a very serious David Brown man, having been brought up with a DB 990 and is, perhaps, just a little embarrassed to own a 'wee grey Fergie' in his collection. However, this tractor is different in so many ways as it has considerable local history and that is so important with such machinery these days.

Fred Thompson from Archill, Newtonstewart, County Tyrone, is still going strong at 93. He purchased the TE-F20 diesel tractor new in 1955 from local dealer George Walson, Newstonstewart. He used it on his 100-acre farm with the milk herd, around the fields on maintenance duties with a carrier box behind repairing the fences and so on. In fact Fred has only ever owned one tractor and this is it, maintenance appears not to have been high on the agenda and eventually it was decommissioned some 15 years or so ago.

One day approximately 10 years ago Fred mentioned he might like to sell the Fergie but told a potential buyer not to call on Friday as he would be in Omagh buying goods for their small shop. Unfortunately, the potential buyer did call on Friday and of course Fred was out. The caller took off the nine-hole drawbar and strap and never came back.

One day Fred called in to see Neville Porter at his trailer-manufacturing works. He was impressed with what he saw and the preserved tractors here; obviously this was the man to have the TE-F20.

An offer not be missed

After visiting the non-running TE-F20, Neville agreed to have it. When he went to offer money for it Fred said no "you rebuild her, that's all I ask" and that's how it came to Castlederg. A friend offered the new owner £300 for the tractor, but the TE-F20 was not for sale.

The engine wasn't seized and after checking the fuel and other things, the tractor was towed down the road and it actually fired up, to everyone's surprise. However, the noise the engine made indicated it was not firing on four cylinders, so an investigation was called for.

With the removal of the cylinder head, another item that was to be eventually discarded, the bores on the 2,092cc Standard 20C unit were well worn. After scraping the huge carbon deposits off the pistons, a hole was found in one of them and the bores were extremely worn and grooved. The new owner was getting rather concerned about the way the whole project looked and his friend said he was glad he hadn't bought the tractor. However Neville felt he just couldn't let Fred down and decided he would fit the restoration job in when time allowed.

An excellent Ferguson mechanic stripped the engine down and replaced the liners and pistons, as this is an extremely difficult engine to work on when it comes to replacing the liners and pistons. New crank shells went in and a second-hand cylinder head came from Neills Vintage from Cookstown – who were to supply many parts along the way for the tractor.

New valves and guides went in the head, which was also skimmed – as were the seats. After fitting new gaskets it was all reassembled. The CAV BPE-type injector pump was overhauled and new filters and injectors were also fitted – the bill on the tractor was already over £1,500, but at least the cooling system was in good order. New water hoses went on and the original wire hose clips were cleaned and re-galvanized and used again. They look wonderful and make the tractor so authentic.

Another serious problem

Gear selection was a problem and with the selector top off it was decided it would be far cheaper to buy a second hand gearbox from Neills at £80 than carry on restoring the original one. Turning to the back end, new hub seals were the order of the day and the brakes were cleaned and refitted. Although it was said the hydraulics would not hold up properly, after refilling the box and checking for leaks it all worked very well.

The steering was good but the axle needed straightening and a new centre pin was inevitable. The side strut arm ball ends were very worn and the axle yoke was similar. These were all built up and ground ➥

On the face of it the TE-F20 looks in good order, but a different story was to unfold.

Even one of the tyres on the rear is the wrong way round and inevitably the top link is missing.

The tractor has been split, the cylinder head is off and the basic strip-down is taking place, but this is only the start.

All now covered in heavy-duty two-pack primer in a warm dust-free environment, and a quick-drying time for the paint – just 15 minutes.

You can see major progress has been made, the modified Sparex exhaust system now looks great.

down and are now perfect, saving more on the cash front too. New 10-28 tyres and tubes went on the rear end and the front rims needed work. Neville looked into purchasing reproduction rims but didn't like what he saw, so built the rims up where required and with some careful lightly-added plastic filler they now look like new again. The front tyres and tubes that came with the tractor live again.

The Sparex underslung exhaust pipe was 'cut and shut' to make a more pleasing fitment and the silencer underwent some modification too, but the exhaust heat shield is still the original one. New Sparex mudguards went on after two-pack primer and lots of rubbing was applied before deciding to use Besa two-pack 'Battleship Grey' (code number 18B25) as the correct colour. It's certainly is a minefield as to the correct colour, but what has been chosen looks about right.

Keeping original

Luckily the original gauges were able to be used and, for the very simple electrics, the dynamo was fine but the starter needed reconditioning and a new fan belt went on.

It was Christmas 2007 when the job of painting the skid unit was undertaken. After sanding down the metal using an air-operated sander with 180-grit pad, two-pack metal primer was applied and left to go hard. This was before applying many coats of heavy-build primer before again using the sander, this time with 300-400 grit pad to make it smooth. Three coats of top coat were applied and the skid unit was complete.

Occasionally Fred Thompson would call in and ask how things were going and really it was his inspiration that kept the project going, as the TE-F20 has been an in-between job. The original bonnet has been used, but with a new Sparex grille and T-piece. In this instance Neville had a top-class concours painter apply the coats to the bonnet in a dust-free environment and the end result has been worth it.

Everything has been kept as original as possible and the restoration has not been overdone.

Fibreglass battery covers have been introduced recently and complete the authentic look.

Decals add an extra dimension to the tractor.

Above: As the original lower links were in a poor state, Neville opted to fit Sparex replacement parts. Finding original arms in good condition is becoming increasingly difficult. The nine-hole drawbar and stays are original Ferguson parts that Neville bought to finish the restoration.

This might be the way to go for other tractor restorers? The bonnet is so important to get perfectly straight and right as it's the first thing ones sees and can make your Ferguson stand out in the crowd and without much extra effort. However, do make sure that the paint you use throughout the tractor comes from the same batch.

When the tractor came out the workshop all spic and span, it looked second-to-none and when Fred went to see it he just couldn't believe what had been achieved – especially as it had new black batteries and authentic covers. Interestingly, one of the batteries is a new heavy-duty 12-volt unit and the other is a dummy.

It cost £4,000 to restore and that's without Neville's time and workshop skill. The end result is something that is still a very genuine machine and in these parts that's worth a considerable amount. There are many TE-F20s about, but restored ones to this standard are few and far between. ■

Fred Thompson, 93, comes to say hello to his old friend, which he bought new 54 years ago.

Finding the cure

Andrew Hall helps to overhaul the rear axle of a TE-F20.

The wheel and wing had been removed to ease access. We found the brake shoes to be contaminated with oil from the rear axle. Note that the tractor was supported on a jack as well as the timber, as axle standss weren't available when the job was started. However, for safety, axle stands are recommended!

While sitting at home one evening, I received a call from an ex-colleague and friend, Bernard, who was working on his Ferguson TE-F20 – it was leaking oil from the left-hand side of the rear axle.

Bernard's tractor had been suffering from poor brakes and he had started to strip it down to replace the brake shoes when he discovered the leak. As he realised there was little point in replacing brakes without remedying the cause, he consulted me for assistance and advice.

On first inspection I observed that his tractor has grease nipples on the axle casings near the brake back plates to lubricate the wheel bearings. On models with these nipples there are two oil seals – one inside the casing to seal the oil in the transmission housing and another outer seal that prevents oil/grease getting on to the brake shoes.

In my experience, leaks of this type are usually caused by failure of the inner seal, which allows oil to run through the bearing and wash into the brakes. A quick cure is to remove the axle half-shaft and change the inner seal, but I felt it would be better to replace the outer seal as well, as it involves little more work.

When I got involved with the job, Bernard had already changed the brake shoes on the right-hand side, albeit with some difficulty – as the supplied shoes lacked some milled slots to enable them to fit. As Bernard has a small workshop of his own, with a milling machine, he successfully machined out the flanges of the shoes to overcome this and the shoes were a good fit.

Access to the seals is not too difficult. Bernard had already removed the rear wheel and the mudguard so this sped the job along.

With the brake shoes removed, the next stage was to undo the ring of bolts around the bearing housing and break the seal between the housing and the brake's back plate. With this done the half-shaft was extracted, together with the remains of the shims and gaskets.

Our decision to replace the outer seal meant we had to employ the skills of somebody with a strong press, as the shaft needs to be pressed off the bearing housing to reach the seal. This always requires a fair degree of force and needs to be well supported, so the axle was sent to Tony Davies, of Chiddingstone in Kent, to do this.

The brake shoes were removed, followed by the six retaining bolts around the axle housing. After a little persuasion the half-shaft was extracted.

The brake plate was swung away to access the inner oil seal. This shot shows the groove with the seal removed.

A view of old and new components shows the damaged shims and old oil seal prior to fitting. The new seal is leather, similar to the old.

A degree of improvisation was called for here. The seal was driven home with a piece of wood, taking care not to distort it.

Gasket sealer was used to ensure a good seal each side of the shims.

A new gasket was fitted in front of the brake back plate. A good seal here is important to prevent brake contamination.

The half-shaft, complete with new outer seal and bearing fitted by Tony Davies, was carefully pushed home and engaged with the bolt heads. Care was needed to prevent damage to the new inner seal from the splines of the shaft.

The new brake shoes were fitted. Note the nuts and washers used to stabilise the shoes. The slots for these had to be milled out as they were absent from the new shoes!

Copper grease was applied to the threads of the brake adjuster on assembly before the brake drum was re-fitted.

The gazebo makes an effective shelter for working where cover is limited. Note the holes in the top centre of the tractor bonnet. These were for the optional lighting system with single headlamp. Bernard hopes to fit a kit in the fullness of time!

Sometimes it is difficult to move the bearing and this proved to be the case, so the bearing had to be cut off the axle shaft. Tony soon had the job done and we were able to refit the axle to the tractor.

The original inner seal was of the leather variety and I was surprised to see that the new replacement was too, as most new seals are made of modern materials such as neoprene. New shims and gaskets were used with gasket sealer, as we didn't want to repeat the job. The brake shoes were assembled and the adjusters freed off and lightly lubricated with copper grease.

In an interesting aside, the tractor had recently undergone a rewire and this highlighted a faulty dynamo, which was replaced with a new one. However, the new dynamo required polarising before use and this was done by connecting the battery and running a wire from the live (non-earth) battery terminal and 'flashing' on to the small field winding terminal on the dynamo. This ensured correct polarity of the dynamo field windings in order to charge the battery.

Bernard's tractor is subject to a continuing restoration as and when he finds time. To overcome inclement weather he has erected a small gazebo over it and it is surprisingly effective in keeping the weather off! ■

My greatest thrill

Barry Dinan shares his thoughts on his favourite tractors.

These two Ferguson TE-20s are both from 1954 and have always worked the land around Gort, County Galway.

It's these tractors that give me the greatest thrill; I enjoy seeing the wear and tear on their hand-worn controls from years of serious hard work.

The one with the trailer and hay on the back is owned by Pat Joe Keane and has worked every day since it was built – and without fail. Pat Joe's 20 has never given any problems through the decades, whether cutting meadows, drawing turf, lifting, scraping or travelling the roads.

He prefers the TE-20 for his work because it's so lightweight and can get him through bogs and swampy fields in all kinds of weather. He rates it well on diesel consumption too, which these days is more important than ever.

Pat Joe got his tractor 40 years back from his friend Frank Cahill – the man with the cap on the other grey Fergie – who supplies him with hay these days. Frank, a good mechanic, uses his Ferguson for all kinds of work around a modernised farm – it earns its place regularly among all the bigger tractors.

So, after the hay has been delivered along with a chat and a cup of tea, it's back off up and over the lovely blue hills of Derrybrien for Pat Joe – a ten-mile journey on his loyal grey Fergie.

"Good luck awhile Pat Joe!" ■

True vintage quality

Dave Beare visits Hubert Thomas – a man passionate about restoring cars, motorcycles… and tractors.

There are modest men in this world who don't brag of their achievements, even to their mates in the pub. Men who are quietly competent and confident in what they do, men who have watched, listened and learned throughout their lives and, as a result, have mastered the skills they need to perform to a level of perfection.

Such a man is Hubert Thomas from Maes-y-Bont in South Wales, an amateur who patiently restores tractors for the love of old machinery, not for profit, and produces standards of finish that are the envy of many professionals.

The Thomas family, like many in rural areas, has a farming background. Hubert's father John worked 140 acres of mixed arable, dairy and pasture – and inevitably used tractors that were available at the time to mechanise activities. John used a Fordson Standard N and a P3-powered version and Hubert had to learn to drive them when he was old enough.

An appreciation and natural ability with machinery soon manifested themselves in Hubert. In the 1970s he started restoring old cars for pleasure – an MG ZB Magnette and Jaguar MkII among them. He has spent much of his life working on farms but never stopped repairing, maintaining and restoring cars, motorcycles – and tractors.

It's not difficult to be mightily impressed by the extraordinary quality of Hubert's work. The word 'quality' is much-abused these days and has been applied to a multitude of products and services which in no way could be described as having any 'quality' worth mentioning, largely thanks to over-use by the advertising industry.

The word is now used to differentiate between the humdrum and exceptional, however, a good dictionary definition is – "quality... a degree or standard of excellence, especially a high standard". A perfect definition for the tractor restorations carried out by Hubert Thomas.

Having finished a TE-D20 to his satisfaction, Hubert had another Ferguson awaiting restoration, this time a 1955 TE-F20 diesel that had spent most of its working life cutting grass at the Pendine Cricket Club, on the edge of Carmarthen Bay. Nearby are the famous Pendine Sands, a very long flat stretch of beach where, on 3 March 1927, Welsh engineer and motor-racing driver J G Parry-Thomas lost his life while attempting to break the World Land Speed Record in his car 'Babs'.

The car was originally built by Count Louis Zborowski using a 27-litre V12 Liberty aero-

Out in the light, Hubert's finished Ferguson TE-F 20 gleams on a winter's day.

engine and final drive was by chains. Parry-Thomas modified 'Babs' considerably in a quest for more speed and was on his final run at over 170mph when the right-hand chain broke, partially decapitating him. None of which has anything to do with Ferguson tractors!

As a result of a lifetime's exposure to salt-laden winds Hubert Thomas's Ferguson was in a bad state, most of it was rusty and it had a lot of internal corrosion, so a total strip-down was necessary. Having split the tractor and removed the engine, work began.

Harry Ferguson had cut a deal with Sir John Black of the Standard Motor Company in 1946 to manufacture Ferguson's TE (Tractor England) design, which was to include a de-tuned version of Standard's 80mm (later 85mm) x 92mm 1,849cc petrol engine destined for use in the new Standard Vanguard car, which was announced in 1947.

However, due to supply problems the American Z-120 Continental overhead-valve engine, first used in TE-20 tractors in 1946, was fitted until July 1948, when Continental and Standard-powered tractors were built alongside each other.

The Ferguson TE-F20 diesel tractor was introduced in March 1951 with an engine modelled on the Standard petrol unit but was, in reality, a brand-new design by Standard in conjunction with Freeman Sanders, using the latter's patented spherical pre-combustion chambers.

This was an indirect-injection engine with dry liners (the petrol unit had wet liners) and was dimensionally slightly larger than the petrol engine as well as nearly 2cwt heavier.

The half-compression lever, situated at the front of the engine where it could be reached while hand-cranking a start.

Considerable modifications were made to the Ferguson's transmission to accommodate the diesel, with the result that petrol engines cannot be substituted for diesels and vice-versa. Output was 25bhp with a modest 17.1:1 compression ratio, two six-volt batteries were fitted either side of the seat and a Ki-gas pump was fitted to aid starting in cold weather.

In case of battery failure Ferguson had thoughtfully incorporated a half-compression control, which could be used if it became necessary to hand-crank the engine into life, not normally an easy job, especially in winter.

Subtle differences

In addition to the engine and transmission, many other minor differences are noticeable between a petrol/tvo TE-D20 and diesel TE-F20. The fuel tank had to be rather crudely modified at the rear to accommodate the Ki-gas reservoir and pump and the new air-filter assembly, with primary and secondary filters, where the battery on petrol Fergusons used to live. ➡

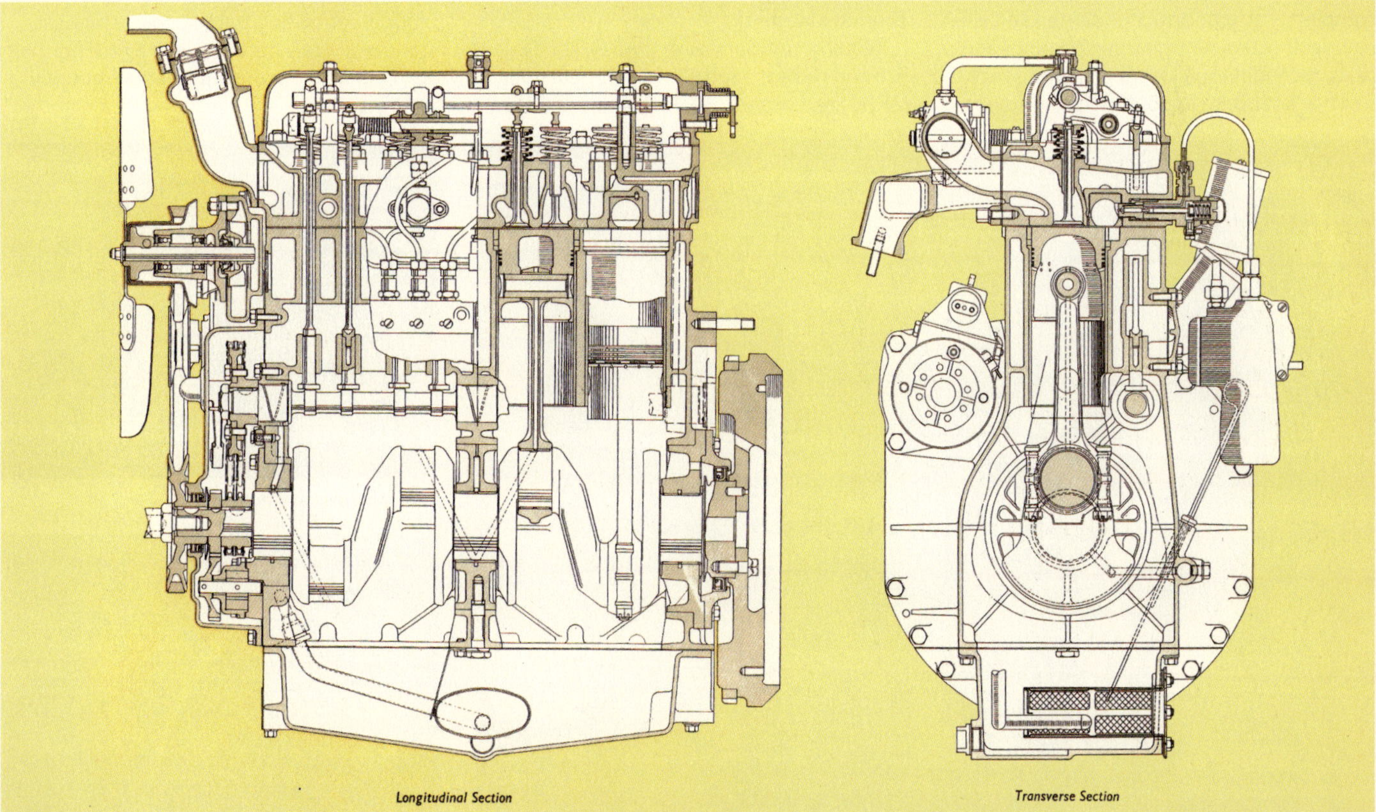

Longitudinal section (left) of Ferguson's first diesel engine from 1951, with a simple 3-bearing crank, and transverse section of engine (right) showing the flat cylinder head with spherical pre-combustion chamber and horizontal fuel injector. A butterfly air-control was fitted in the inlet manifold, but this was abandoned on later Ferguson tractors.

Filters seem to have become an obsession with Ferguson, as two fuel filters were used as well as a filter on the crankcase breather – perhaps the Banner Lane plant was more aware of the very dusty conditions many of the exported tractors had to work in.

Hubert's TE-F20 diesel was completely rebuilt from top to bottom, with new pistons and liners, reground crank, new bearings, all-new valves, guides – everything. "It's not worth doing half-measures at this stage, you'll only regret it later when you have to take the engine out – again" he says.

"Accurate injection-pump timing and a reconditioned pump are essential for easy starting" – proven by the fact that this tractor starts from cold without Ki-gas assistance in just a turn or two.

Hubert, now 73-years old, maintains there's no 'secret formula' for a perfect restoration; "It's basically many, many hours of hard work and elbow grease. There aren't any short-cuts, they always show in the end."

Hard work and attention to detail; even seemingly unimportant items are not only fully restored to better-than-original condition but given that extra burnish that shows real care.

For instance, though a TE-F20 never left Banner Lane like that, giving the cast-alloy end casing of the dynamo a going over with the polishing mop means it looks just great. A better finish, but without going over-the-top.

A lot of extra plumbing – the injection pump, injectors and associated pipework take up a lot of space, note twin fuel-filters & crankcase breather filter.

A job well done

The entire tractor was methodically restored without sparing the time or expense of doing the job properly; the hydraulics were completely stripped, reconditioned and all-new seals fitted, new gearbox bearings and seals, reconditioned steering-gear, correct-size black rubber-case batteries and covers, all-new (expensive) Goodyear tyres, new cooling hoses, remade injection pipework – the list is endless.

Adjoining Hubert's workshop is his tractor house, a revelation – the building itself is an old wooden school class-room that was bought in Peacehaven, East Sussex, dismantled and brought all the way back to South Wales, then re-erected in the Thomas's yard.

Keeping company with the TE-D20, TE-F20 and Hubert's son Eifion's four-cylinder MF 35 are a superb, original Vauxhall Cresta and Triumph 350 motorcycle, all enjoying the luxury of fully-carpeted and heated accommodation!

As Hubert Thomas says, there are no short cuts to doing a proper job and they always show, his attention to detail is exceptional. It was a great pleasure to visit the Thomas family, to talk to true, but modest, enthusiasts who take endless pains in rebuilding tractors to the highest standard, rarely have we seen a final finish of such quality – for once this term is entirely justified. ■

Roughly cut-away fuel tank to accommodate larger air-filter and Ki-gas equipment. Note that the paint finish is exemplary.

Everything as it should be!

A lot of cable conduit needed for the batteries.

The flying Ferguson

Sometimes finding a gem like this is a question of being in the right place at the right time. Well, Brian Colley was in that place, much to the envy of the rest of us!

When Brian Colley went, as any keen collector regularly would, to his local H J Pugh sale, at Southam in Warwickshire on 17 November 2007, he was not expecting to find anything special – but just look at what he came away with!

He has since taken the tractor to local shows, to Tractor World in Malvern and to The Great Dorset Steam Fair, where it attracted a great deal of interest.

The tractor is serial number 311731, which means it is a 1953 model. We believe that it originated from the Peterborough area, but have no other information on the tractor.

Investigating the tractor's military origins through its RAF number has unfortunately not helped uncover any further information. However, our investigation did lead to some interesting findings. While the David Brown Taskmater and Tugmaster tractors were commonly used for towing aircraft, we have unearthed photographic evidence of TE-A20s performing this task on NATO airfields.

This tractor, fitted with grass tyres, was almost certainly used just for grass cutting. The number on the yellow bridge disc ('C 10') is a bit mystifying. Normally this would indicate the weight of the vehicle in tons as an indicator of its ability to cross temporary military bridges – but this practice had almost certainly ceased on RAF vehicles after the war. Besides this, the tractor certainly does not weigh 10 tons!

During the war, vehicles were marked to indicate their command and unit so perhaps this 'C' indicates Coastal Command. This marking would generally be put on the wing in white lettering, but it is possible that a bridge disc was used because the tractor has no wings! ■

Brian Colley on his Ferguson TE-F20, resplendent in its RAF livery.

Harry Mustey is shown with some of his collection. In the background, his son Graham is shown on the Fergie P6 conversion, together with the 1953 TE-F20 and the 1948 TE-A20.

Harry's converted

Is it a first? Angela Walker finds a Ferguson P6 conversion from scratch.

For Harry Mustey, of Lymington in Hampshire, tractors have always been part of his life and retirement has meant more time to spend on his love of restoration. Above all, it has been an opportunity to realise his long-held ambition to build a Ferguson P6 conversion from parts.

Harry grew up on a dairy farm north of Andover, where he remembers helping his father look after the horses. Later, when these were replaced by tractors, the family used mostly Fordsons, including Standards and an E27N.

Moving to the Lymington area when he married in 1960, Harry worked as a maintenance engineer at a research station owned by what was then the Ministry of Agriculture, Fisheries & Food.

By 1973 he had decided he needed a tractor to maintain his acre of land so he bought a 1953 four-cylinder Ferguson TE-F20 from Savill Tractors, of Salisbury, which he still owns. After giving it a coat of paint, he started entering it in shows.

Two years later he added a 1948 Ferguson TE-A20 with a P3 Perkins three-cylinder diesel engine to his collection. Originally fitted with a loader, it had been used for landfill work and was in a very sad state, so Harry set about completely rebuilding the engine. Also, because the TE-A20 was designed before the pto-driven rotovator came into use, it was fitted with a Howard reduction gearbox, a 4:1 reduction box which reduced the tractor's ground speed, yet kept the revs up high enough to run a rotovator on the back.

Harry decided to take the gear apart and reverse it, so it could go four times faster. With a comfortable cruising speed of 30-40mph, he could drive the tractor to shows. His son Graham, too, has driven it in the Bournemouth-Bath commercial road run, as well as to Newbury and the Yeovil Festival of Transport.

This early picture of the P6 engine shows the extended trunion bar and drag links. Also visible are the six-bladed fan and the water pump in its lowered position.

Harry had the idea of building a Ferguson P6 conversion literally from pieces. Beginning the project in 1993, he set about the painstaking task of sourcing the parts. As Graham points out, Harry likes a tractor to be as completely original as possible and he used only original Perkins and Ferguson parts from the 1950s for this project.

The first step was to buy the Perkins P6 diesel engine, the first of the iconic P series launched in 1938 by Frank Perkins Ltd of Peterborough. Harry was lucky enough to find one locally that had been used in a recovery truck made by Guy, the British company formerly based in Wolverhampton.

The truck engine is up-rated with aluminium pistons – normally the P6 tractor engine would be 64hp, but originally being a truck engine, this one is 72hp and runs at higher revs. It was decided to mate this to a Ferguson FE-35 transmission and rear axle because of its high-low range, six-speed gearbox and heavy-duty pto, as well as better brakes.

Harry is the first to admit he could not have done the project without the help of many friends, and he is grateful to them. For example, when he needed to obtain, re-machine and fit the engine back plate with a heavy-duty flywheel, friends were on-hand to help with machining the flywheel and dual clutch assembly. The double clutch was necessary, he explained, to keep the pto separate from the land drive, so that when he dipped the clutch to stop the wheels turning, the machinery did not stop as well.

He had to strengthen the sump to support the front axle carrier, whereas other conversions would have side rails to take the extra weight. To overcome this problem, Harry built a sump out of two suitable ones, which he modified and welded together.

The timing casing also had to be changed to lower and modify the water pump, so that the fan was lined up with the radiator. A replacement six-blade fan was necessary to cool the bigger engine.

Other modifications were needed – Harry had to extend the trunion bars and the drag links. The steering blocks had to be blocked and reset and the dash panel lifted slightly. Harry's friends were on-hand to obtain exhaust manifolds and to modify the inlet manifolds.

The fuel tank had to be hot-pressed to fit over the engine rocker cover. The bonnet was extended but was not the simplest of exercises.

Harry pointed out that the Fergie bonnet is not very parallel, so is very difficult to extend and make it look right – "What I had to do was cut sort of teeth in the metalwork and force them together, then rivet them, finally welding it and finishing it." ➥

As the Perkins P6 engine was originally from a truck, it is uprated at 72hp with aluminium pistons.

Harry's son Graham demonstrates the Fergie P6 conversion. Graham helped his dad with the project and describes tractors as "very much a family thing."

Oversize tyres had to be fitted to the front and rear to be compatible with the size of the tractor and to do this, Harry had to lift the mudguards.

Eventually it finished up only ten inches longer than the normal wheelbase, so that avoided turning drift on slippery ground.

The work took more than ten years, although Harry also completed several other restorations during this time, such as a puller, a couple of classic cars and a couple of mini tractors. As Graham put it: "He's always got something on the go."

Graham, an agricultural contractor who also lives near Lymington, was there to help out if needed, such as when fitting the starter ring gear and clutch assembly to the gearbox bell housing, which Harry remembers as the most difficult part of the whole operation.

The years of painstaking work had their reward in 2006 when the tractor was finally completed. As far as Harry and Graham are aware, a Ferguson P6 conversion has never before been built from parts. Now it can be seen at steam fairs and rallies – Harry's tractors are never kept as show pieces, but are regularly taken out for ploughing matches, road runs and all kinds of events. ■

Harry uses his Italian-built Massey Ferguson 154C for ploughing matches. Unusually for a crawler, it has three-point linkage.

Harry's man to the core

Peter Harris has a disability, but that's not going to stop him building his Ferguson collection, as Kim Jackson discovers.

Peter Harris is a dairy farmer's son from Ellerdine, near Telford in Shropshire. He has always been interested in machinery and started off by restoring horse ploughs and clothes mangles until, with encouragement from his mum and dad, he progressed to a tractor in 1994.

Peter's disability means there are some tasks he can't undertake, such as using some power tools, so he either uses the manual equivalents (he is a master of the wire brush) or somebody helps him. In the early days it was his grandmother who was his main aide and her death in 1996 left a big hole in his life.

The first tractor Peter restored was a 1954 petrol/tvo-powered Ferguson TE-D20, which came from a market garden in the village; Alfred Innes had owned it from new.

Fergusons had always been on the Harris's farm – currently there is a Massey Ferguson 135 (the scraper tractor), an MF 148 and an MF 362 that has a loader attached to it (most field work is carried out by contractors) – and Peter was happy with the choice he made of the tractor he would restore.

Next was a David Brown Cropmaster, which he managed to renovate by 1997, although he always felt it was a mistake getting this Yorkshire tractor and never really enjoyed doing it up. The mechanics were sound and it received mainly cosmetic work from Peter, who sold it shortly after finishing it.

His third project was a Perkins P3-powered TE-20, and he ended up restoring this one twice because he was not happy with things first time round. This tractor again was sold to make way for better things.

Peter's third Ferguson is still part of his collection, a Shropshire-registered 1948 TE-A20 with an aluminium bonnet, which he bought in good condition through an advertisement in the local paper. Basically all he did was repaint it.

Up to that point, it could be argued that Peter's collection of three Fergies and a Cropmaster lacked interest, but that was to change. He has a great attraction to grey and gold Ferguson FE-35s – "most people hate them, but I love them," he says.

His first search led him, through the classifieds in *Tractor & Machinery*, to Shaftesbury in Dorset as Richard Pocock had a 1957 diesel-engined FE-35 for sale. Richard sent Peter a photograph, from which he purchased it, and it arrived the week after Christmas 2001. Peter's first grey and gold restoration took him six months with help from a new friend, Keith Fenn, and then it was time to look for another tractor.

A visitor to the Harris's farm told them about a petrol/tvo-powered Ferguson ➡

Peter calls George Lloyd his 'best mate/apprentice' and George is operating the petrol tractor which he spent a lot of time with Peter restoring. In the background are the Harris's Jersey cows.

Three of a kind. All the tractors are equipped with Ferguson pick-up hitches, like the one on David's petrol/ tvo-powered example. Peter spent a lot of time searching out enough hitches to fit on to the entire fleet.

FE-35 on a smallholding at Ashbourne in Derbyshire. Peter duly went to look and found it at the bottom of an orchard in a very bad way. Although the tractor no doubt left Banner Lane in its grey and gold livery, it had subsequently been painted purple.

As most people do, Peter lifted the bonnet to look underneath, only for it to fall apart! The manifold had broken off, but at the bottom of the U bend, so water had fortunately not run into the cylinders, the engine still turned and a deal was struck. The tractor arrived at Ellerdine and work started.

Keith wasn't impressed at all with the amount that needed doing, but still helped out where he was needed. It took seven months to restore and Peter commented: "I wouldn't like to do another one quite as bad as that again."

There were now two grey and gold Ferguson FE-35s at Ellerdine and Peter went to the *Tractor & Machinery* Show at Rochdale in November 2004, where he asked Sam Neill from Northern Ireland for help in sourcing a straight petrol Ferguson FE-35. Sam wasn't very hopeful but suggested he might have more luck finding one elsewhere in Europe. In the end Peter did find one for sale in the UK, belonging to Rob Archer, and bought it in March 2006.

Like the petrol/tvo-powered FE-35, this tractor was not grey and gold. It was probably once council-owned, as in places it had green and yellow paint on it. But underneath all this, gold could be seen and Peter set about removing the old paint with a chisel and wire brush.

His collection also features a rather-rare petrol-powered Massey Ferguson 35. Between 1999 and 2001 Peter studied for a National Diploma in Agriculture at his local agricultural college, Walford, where there was an MF 35 tractor that had been bought new by the college, but was no longer used.

It was parked outside the workshop and had been stripped down by two people who then lost interest in it. Peter was asked if he would like to restore the tractor as part of his diploma and for the next two years he spent two mornings a week working on it. The college who, of course, owned it paid for the new parts and materials needed for the tractor.

At about the time Peter finished at Walford the surplus tractors and machinery were put up for sale, so Peter put in a tender for the MF 35 and it is now back at Ellerdine.

The collection is continually evolving and includes a good assortment of Ferguson implements and accessories such as one

of Peter's early projects, a single-furrow reversible 'butterfly' plough. From this he graduated to doing up a two-furrow plough and he was rather pleased that he got in at the right time as now they seem to have quadrupled in price.

One of the implements of which Peter is most proud is a very smart Ferguson three-ton tipping trailer. The once redundant and worn-out trailer was acquired from a farmer from the next village, Bill Preston of Rowton, on the understanding that "if I could restore it, he would give it to me."

The first thing Peter did was strip everything back. Next, new cross members were laid in the chassis and all new timber was sourced for the floor. Peter cut it himself with a normal hand saw at the rate of two boards a day and it took him a fortnight. He then needed help bolting it as a drill was needed. The back boards of the trailer were made from scratch.

His other implements include a transport box, earth scoop, potato planter and a weeder which, when Peter acquired it, was "very, very rough." A steerage hoe came from another local farmer with the same condition as the trailer that, "if I could restore it, he would give it to me."

A cultivator came from the same sale at Walford as the Massey Ferguson 35. There are also a mower, discs and a fertiliser spinner from the later Massey Ferguson era. The mower came in a job lot – this one and an incomplete mower – from a Hall's collective sale held at Montford Bridge airfield outside Shrewsbury. Peter restored it and sold the incomplete mower.

Peter has been fortunate in that the tractors he has acquired for his collection have not required major mechanical repairs to engines, gearboxes or final drives. He readily acknowledges the help from many friends and family who undertake jobs involving power tools and other work he is unable to do and wants to thank his mum and dad, Keith Fenn, George Lloyd, Jim Hinks, Neil Lewis and many more people for their help, support and dedication throughout the years.

But Peter will always undertake as much of the restoration as possible. Tasks like wire brushing, painting and cutting the boards for the trailer took him a lot of time and effort to complete, but he would never give up.

Peter is a Ferguson man to the core and when he is not with his tractors and implements he is quite likely to be on the internet looking for his next project. ■

Maurice takes the FE-35 out into the sunlight.

The beet goes on

Peter Love visits Maurice Houghton in Cambridgeshire to see his excellent restoration of a 1957 Ferguson FE-35.

Maurice Houghton spent all his working life with F A Standen Ltd, starting at their St Ives, Huntingdon branch where he stayed for some 49 years and 8 months, then the last four months before his retirement at the Ely branch. Three generations of his family were to work for this company, which is now called Thurlow Nunn.

Having worked mostly on Ferguson and classic Massey Ferguson equipment, his knowledge of these tractors and their parts is invaluable to all preservationists today.

While out on call one day, Maurice came across a Standen Solobeet harvester that had been converted into a sprayer, standing in the bushes at Broughton, Cambridgeshire. As he serviced various machines in the area he always kept his eye on it, as, having previously had a close look underneath the yellow paint, he knew that the skid unit was a 1957 grey and gold Ferguson FE-35.

Then, in 1999, farmer David Castle – a friend of the Houghton family – kindly agreed to part with the machine. Now with the help of Maurice's son Gary, a keen motorcycle collector and another 'ace' mechanic, they cleared all the brambles to get to it. They moved the outfit to Colne, were it was lifted off the transporter with a telehandler.

That was in October 1999 and it was intended to have the tractor restored in just 18 months. In the end it would be an eight-year job, but well worth it. As Maurice was in the 'trade' he collected lots of the original parts that are needed for such a project as this. These came in very handy indeed and saved lots of stress along the way.

Out intrepid Ferguson mechanic says it's always best to get the tractor running first so you know where you are and you can programme the project ahead. With the battery leads attached the engine was wound over, but no go at all. The four-cylinder indirect-injection diesel engine was checked out and it was found that the injector pump was seized and the injectors were leaking too.

After this was sorted out, and refitted, the tractor eventually fired up, but no oil pressure! So a major strip down was the order of the day – at least Maurice knew what he was working to.

The 23C engine was stripped right down to reveal that the crankshaft and shells were worn, but the pistons and liners were in very good shape indeed. It was obvious these had been overhauled not long before the outfit suffered problems and was decommissioned. The timing chain was removed and it was noticed that the chain and adjuster were very worn and had slipped, hence part of the poor starting. The oil feed pipe was badly worn as well, so this was sorted with a second-hand part. On further inspection the Ricardo-designed cylinder head was in good order.

The oil pump was stripped and the gears were found to be in good order. It was cleaned and tested and ready to roll once more.

New crank seals were fitted and with the new timing chain and tensioner in place, it was all systems go.

It might sound very easy, but these engines are not the best to work on and particular effort has to be placed on getting the timing right. As is well-known, the 23C engine is not a good starter in the first place, that is why you need everything in good order from the outset.

With the block and head all back together and the radiator in good order, new hoses and a new water pump were fitted. A replacement fan blade was bolted on, as was a new air cleaner. Just like the injector pump, all the diesel filters and pipes were replaced or cleaned and a methodical approach was taken to have it right first time.

The objective is not to constantly be going back and repairing things, as does happen. All the copper washers were replaced and Maurice made sure they were well annealed and the right diameter. This is something to think about with so many metric sizes around, remembering that this tractor was built at a time when imperial was still the standard.

The engine was then run up and Maurice was happy with the way it had shaped up. He was to keep a log of where, when and how, which was a handy reference particularly as this turned into an eight-year restoration.

The starter motor and dynamo are the original units, which have been overhauled and painted, two new 6-volt 'black' batteries were obtained and a new battery frame was made. All the electrics were overhauled and the correct terminals used, which makes such a difference when looking at the tractor for period correctness. A replacement ammeter, oil gauge and tractormeter had to be fitted. ➡

Above and below: This is what Maurice saw when he first went to the farm, but he knew the Solobeet harvester (fitted with a Winsam cab) was something extra special.

Lots of enthusiasm is needed for a restoration such as this. The straw and twigs were removed from the air cleaner area as seen in the picture.

The working project gets a road-test. With the engine now fully overhauled, Maurice gets ready for the next stage.

The gearbox and back axle were in excellent order and, after being cleaned up, were refilled with oil. The hydraulics were good, but the levelling box and lever had to be sorted out. The box bearings had collapsed inside, but luckily it's not a very difficult job to replace them. The levelling lever was all sorted out on the quadrant as well.

Interestingly enough, one of the top hydraulic link arms still had the original gold paint on it and the other was in red oxide underneath and had never been painted properly. The brake shoes were replaced and the linkages were freed off using copper slip around the pivot pins.

The front axle was overhauled with just the centre pin replaced, the king pins were excellent and the wheel bearings were cleaned up, degreased and readjusted.

All was good with the steering box and it just needed the top bearing adjusted. The steering joint rubbers were replaced as a number were missing and the steering wheel, the original, cleaned up remarkably well and looks like new.

While this was going on, Maurice started hunting for the correct footplates as the pattern ones he looked at didn't have the holes in the correct places. He hunted high and low in breakers yards and eventually came up with one, which matched the 'old' stock item that he turned up for the other side.

In regard to the mudguards, our intrepid owner was able to use the frames again and new skins were fitted over the top; all-in-all a very good job was done.

The bonnet was very straight and could be used again, but the grille panel had to be replaced, and the large fixing screws of the correct sort were manufactured by a friend. In fact, at the time of my visit, these were away being chrome plated, as on the original tractor.

The front lower panel was also replaced and the starting handle hole was drilled and punched out, as it is not manufactured on the panel today.

The 'white' exhaust was also renewed and black heat resistant paint was applied to the exhaust manifold. A new Ferguson badge on the grille had to be fitted as the old one was beyond use, but the badges on the side (near the steering wheel) of the bonnet are original.

The tools and parts book that were supplied new with the FE-35 – still one spanner to find!

Maurice is a firm believer of lightly sandblasting the tractor. This is the best way to make sure that rust particles are all disposed of before painting can commence.

An FE-35 is an ideal tractor to restore in a normal 'home' garage and can be worth more than the earlier TE-20 range.

It is essential to make sure that the manifold heater plug is working as these engines are not the best of starters.

After bedding in the steering do check that the track rod end nyloc nuts are all tight.

Little wear was found in the brake linkage and pedal. However, a new pivot pin was fitted as a precautionary measure.

Make sure that any diesel leaks are stopped before painting your object of beauty, or it could cost dearly later on. ➥

Maurice on the FE-35.

A cover over the pto shaft and drawbar was also obtained from the spares cabinet and restored and painted in the grey livery. These were then fitted and they certainly stand out against the gold of the differential housing.

Maurice had a set of four wheels for the tractor of the correct type and these were complemented with a brand new set of Goodyear tyres.

Time was getting on and the project had been going for just over six years when the tractor was prepared for the sand blast of the major units and parts. Every hole, nook and cranny was covered up, to make sure that the sand would not get in to any component at all.

The painting job came out well, with Charles Freitag on the spray gun. After the sand blast and clean up, a zinc coat was applied to the raw metal, followed by red oxide, followed by three coats of Sparex Agri Gold (S84097).

Maurice looked at various interpretations of 'Ferguson Gold' by three manufactures and decided on the Sparex version. He feels this is the right shade.

Before painting, the panels had been bolted to the tractor to make sure they fitted properly. After aligning them, they were removed again for painting in the light grey Ferguson livery. After leaving the paint to go hard on the panels, it was assembly time. All the nuts and bolts of the correct type (not metric) were hand painted with a children's art brush and not just left as cadmium plated as some people do these days.

Maurice obtained the correct stabiliser arms for the year of tractor and fitted these on the rear end, adding a nice touch to the restoration. The starter, manifold, and dynamo were all fitted along with the new fan belt. The dashboard was refitted with the new gauges. Maurice experienced problems with the rubber ring around the ammeter, so the original one was used. Unfortunately this has seen better days.

The fuel tank had been pressure washed out and refitted; making sure that it was all good inside. The water was then replaced in the radiator, mixed with four pints of anti-freeze for safety.

Maurice was so lucky to be able to obtain a Lucas 302 tractor lighting set thanks to Tony Macer's efforts. It didn't have all the items in the kit but was a very good start indeed. Missing was the number plate backing plate and rear lights, besides one or two other items. However a pair of original style lamps was bought and fitted and away Maurice went with his fitting instructions. A 100 metre length of the correct three-core cable was obtained at vast expense.

All went well and after one or two adjustments the tractor was ready for its debut at a local show. It was admired by so many and created quite a stir, but didn't walk away with any prizes.

This was a disappointment to some, but as Maurice's says it's the taking part that counts and he was well pleased with the comments people made about the tractor. As he says it's the attention to detail that makes a tractor stand out from the rest and that is certainly the case here.

So what's next? Well, as a born and bred Massey Ferguson man he is already excited about his new project – a 1971 Massey Ferguson 165, which breathes very heavily for a start, but essentially is a straight tractor underneath. ■

You might think that the number plate backing plate is just flat, but you can see the correct one is not.

A pto cover had to be found for the tractor, and painted in the grey livery. Some people paint them gold by mistake. Note the correct metal straps around the rear axle to hold the wiring in place towards the rear of the console along with the power take-off engage and disengage lever.

The Lucas lighting set was an added bonus to the restoration, but making sure it was accurately fitted and finding the correct wiring terminals took some doing. Note that wonderful battery cover, these are obtainable.

The headlights have the correct Ferguson glasses in them, something that needs to be looked for in a restoration of this kind.

The draft control lever with all the correct writing, something not covered up with filler paint.

Secondhand rims were obtained for the 35 and the Goodyear tyres are the correct type, as fitted to tractors in 1958.

Graham Forster on his Ferguson FE-35.

The show goes on

Restoring a Ferguson FE-35 and running a show. Bob Weir reports.

Organising a vintage tractor rally is not for the faint-hearted, as many people have found out to their cost. But when Graham Forster's daughter suggested the idea after visiting the Festival of Transport at Llandudno, Dad decided to give it a whirl.

Graham is a well-known businessman in the Carlisle area and his background is in farming. He then ran a medium-size haulage fleet and now owns a builders merchants near Longtown.

He inherited his love of tractors from his father and still looks after the family's original Ferguson FE-35.

He said: "My father used to farm a couple of smallholdings and kept beef and dairy cattle. He took delivery of the FE-35 on 4 August 1957, at a cost of £610."

The FE-35 was launched at the Smithfield Show in 1956. The tractor had been preceded by the American-built TO-35, which had a similar specification. It was originally designed before the merger with Massey-Harris in 1953, and was a significant upgrade on the TE-20.

The new tractor boasted more power, better hydraulics, and was available in petrol, petrol/tvo or diesel format. Other improvements included independent brakes and a dual lever control for the hydraulic lift. Deluxe versions were also equipped with a live power take-off unit, which ran independently of forward progress and was ideal for tasks like baling, root crop harvesting and muck spreading.

The colour scheme of gold and grey attracted a lot of interest and the new Fergie soon acquired several nicknames, including Golden Belly and Golden Bird. In 1957 the distinctive livery was replaced by the red and grey livery and triple triangle badge of Massey Ferguson.

Graham's tractor (serial no: SDM 35610) is the diesel version, fitted with a dual clutch.

He said: "My father always promised that when the tractor had finished its working days, he would give it to me as a present. By the time I got hold of it, it needed a full 'from the bottom up' restoration, which I carried out myself."

Graham also remembers that he had a particular problem matching the Fergie's gold paint.

"The tractor had been bought from Fewsters of Hexham, the local Ferguson dealership," he recalls. "One day I got a phone call out of the blue from the original salesman who had arranged the deal. He was getting on a bit by this stage but had somehow heard that I was about to start the restoration and was ringing to make sure that I used the right shade of gold paint."

Despite the fact that the tractor was covered in rust, Graham found a trace of gold paint lurking behind the battery box. The veteran salesman confirmed his suspicion that the correct shade was Jaguar Sandstone.

He said: "He also mentioned that he had recently been shown another FE-35 restoration, but that the owner had used gloss paint in error. He wanted to remind me that to be truly accurate, the tractor should be finished in matt."

But as things turned out, this was easier said than done.

"I popped along to Carlisle to see my local paint specialist," said Graham. "Although he was familiar with Jaguar Sandstone, he warned me that the modern mixes wouldn't accept the matting agent."

The specialist then brought out an ancient cardboard box of colour chips and, after some rummaging around, unearthed an old shade strip that had once been popular on Vauxhall cars. It was also a dead ringer for Jaguar Sandstone.

Graham said: "More importantly, there was a formula to remind the would-be restorer to mix the paint in two-pack, as opposed to the modern cellulose. This allowed it to accept the matting agent."

Graham has been a regular on the vintage vehicle show circuit for many years. Once he had finished the restoration, he started taking the Ferguson to rallies accompanied by his elder daughter Mel.

Graham's 1952 Ferguson FE-35 respelendent in its grey and gold colour scheme.

"The only problem was the travelling," he recalls. "Apart from a couple of local shows, Carlisle didn't have much to offer the classic vehicle enthusiast and this usually meant a lot of driving. We were stuck in a traffic jam on our way home from the 2008 May Bank Holiday event in Llandudno, when Mel had her flash of inspiration."

The majority of people would have wisely nipped the idea in the bud. Yet the more father and daughter thought about the concept of a Carlisle rally, the more it seemed to make sense.

"Carlisle is the centre of a large catchment area, and was crying out for a major show," said Graham. "I decided to have a word with my friends on the local council to see if we could use Rickerby Park, which is located just off the city centre."

This was the beginning of a hectic 12-month period, which would eventually see the birth of one of the best new rallies on the UK circuit.

Graham said: "I realised that we were only going to get one shot at organising the rally and we had to get it right first time. Although I was keen to invite as many tractor enthusiasts as possible, our first step was to get the support of the local steam engine fraternity. We needed a wide audience to increase our chances of success, and steam engines have always been a big favourite with the public."

Father and daughter then spent the rest of the summer attending weekend rallies and handing out thousands of leaflets. They acquired a couple of databases from other clubs and, with the help of a newly- ➥

Left-hand view of Graham's FE-35.

designed computer program, Mel put together their own list of contacts and exhibitors, and started constructing their Border City Steam Fair website.

Apart from information about their own rally, Mel also included a comprehensive 'what's on' list for the Cumbria area to whet the appetite of potential visitors.

Graham said: "Having got our list of exhibitors, we then had to attract the interest of show goers. This obviously cost a bob or two, but I managed to find the money from our own resources."

Mel printed some posters and her army of family volunteers spent many evenings criss-crossing the countryside to spread the word.

"We would get in our cars and head off in opposite directions until we had pasted a leaflet on every village notice board for miles around," Graham recalls. "We then erected larger posters on all the roads leading in and out of Carlisle. Fortunately, the time spent turned out to be worthwhile as the initial response exceeded our expectations."

Graham already had an advertising contract for his builders merchant business with a local radio station and during the run up to the Whit Bank Holiday event he pulled a few strings and substituted his regular message with a jingle about the rally. He also dug deep into his pocket and in the final few days leading up to the event splashed out on some advertising with Border Television.

Mel had already started posting out applications in February 2009. They were being returned at such a rate that, barring unforeseen mishaps, the rally's success seemed assured.

Graham said: "We had agreed from the outset that we were going to donate any proceeds to the Great North Air Ambulance service, the Eden Valley and Eden House Children's Hospices."

To minimise the workload Graham decided to dispense with programmes and judging was also excluded. To maximise attendance, late entrants were allowed to get in on the day.

He said: "The only thing that remained was to have meetings with the police, health and safety and other officials to make sure things ran smoothly. We even had to take out barman certificates to allow us to work behind the beer pumps. Fortunately, everyone involved realised the benefit to the city, and was very supportive."

Holding a three-day jamboree also meant organising a lot of entertainment for what promised to be a bumper crowd.

"Once again we decided to split the workload," Graham explained. "I organised the marquees and beer tent and Mel sorted out the disco. Apart from taking out insurance, the only thing we couldn't make an allowance for was the weather."

Fortunately, the elements were on their side and the Border City Steam Fair was a resounding success. For further information visit the website: www.bordercitysteamfair.co.uk ■

The Standard 23C diesel offered more grunt than the earlier 20C-powered TE-F20.

The original, straight-edged, blue Ferguson badge of the TE-20 was updated to suit the new FE-35.

All of the tractor's original accessories have been meticulously restored.

What an autumn scene!

The statue has been placed right in front of Harry's home farm at Growell.

Harry is home

A fitting tribute to Harry Ferguson has finally been unveiled. Gary Connolly was there.

One of the things (according to my overseas friends) that sets people from Britain and Ireland apart from others is our fascination with the weather. It's regularly the opening gambit in a casual conversation when locals meet and most reports make some reference to the climatic conditions.

I've had the good fortune to cover some vintage gatherings in excellent weather and there have been other days when it has been less than kind. In my time as a reporter however, I have never experienced weather conditions as bad as those endured at the opening of the new Harry Ferguson Memorial Garden outside Hillsborough in Co Down.

The local news headlines gave some indication of the uniqueness of the rainfall – "Over 60mm of rain in 24 hours"... "New Belfast underpass under 20 feet of water".... "Three-quarters of August rainfall in a single day".

Remarkably, despite all of this, the opening of the Memorial Garden took place as planned with Harry's granddaughter, Sally Fleming, performing the official unveiling of the great man's new statue at 3pm as planned.

It's been a long hard road for the Harry Ferguson Celebration Committee and it seems unfair that, after so many years of tireless fundraising, their big day was to be blighted by the rainfall. Regardless, all involved can be rightly proud of what they have achieved.

It was back in 1999 that a number of Ferguson enthusiasts held a rally, which was subsequently referred to as the Harry Ferguson rally given its location on the farm that had been the man's birthplace.

At the time, the only reference on the site to the Ulster inventor was the plaque on the wall of the farmhouse referring to his birth; those behind the rally resolved to pay a more significant tribute to one of Ulster's greatest engineers.

It quickly became apparent that a six-figure sum would be required to realise the group's ambitions and so the aforementioned committee was formed and fundraising began in earnest. Rallies were held on the site each year and vigorous fundraising had the dual benefit of raising money and promoting the cause.

It was through one of these fundraising activities that the Midlands Branch of Friends of Ferguson Heritage, based in Virginia, Co Cavan, became aware of the efforts of the team from Ulster. The Midland Committee decided to put their collective shoulder to the wheel.

A TE-20 tractor was donated to the Midlands group and this tractor became the focus of the fundraising activities of this committee. The tractor is believed to be the first to have travelled the length and breadth of Ireland (from Mizen Head, Cork to Malin Head, Donegal) in 2001 and from Spiddel, Galway to Cooley, Louth in 2003.

Funds raised from these and other events were contributed to the memorial fund and since 2003 the two committees have worked together in a spirit of cross-border co-operation to realise their common aim of a lasting memorial to Harry at his birthplace.

If Harry was alive today I'm sure, given the various accounts of his early life, that he'd have mixed feelings about the farm at Growell.

As on all farms in the early 20th century, work was hard and mechanisation was almost non-existent. It's been said that "necessity is the mother of invention" and there's no doubt in my mind that the drudgery associated with some of the mundane farm tasks led Harry to believe that there must be a better way.

So difficult was the work at home that Harry gave serious consideration to emigrating to Canada and it was only the offer of a mechanical apprenticeship with his elder brother Joe in a car dealership in Belfast that encouraged him to stay. Who knows what path tractor development would have taken had Harry gone to Canada as he had planned?

Regardless of the hard work, Growell was the place where Harry grew up with his family of ten other brothers and sisters and after his departure he often returned to visit, maintaining his association with the wider family and the locality. It's fitting therefore that the primary lasting memorial to Harry in the land of his birth is on the farm where he grew up and gained his inspiration for many of his inventions.

The committee had planned that the unveiling of Harry's statue was to be the culmination of a celebration rally. The weather obviously put a damper on attendance and ground conditions were difficult, but it was encouraging to see the number of enthusiasts who had made the effort to turn up regardless.

Big Percy Donaghy was there with his MF 4840, certainly the biggest Ferguson derivative on display, though at least he had the advantage of a spacious cab that kept out the rain.

I also caught up with David Graham, local restorer, who remarkably had arrived on site in the early hours of Saturday morning having left directly after the annual Enniskillen town road run the previous night.

Other local Ferguson men that I met on site included Jim White, the well-known collector from Antrim and Sam Neill, an acknowledged local Ferguson expert who had provided much assistance to the Harry Ferguson Celebration Committee. There were many others who had travelled from the Republic of Ireland, Scotland and England to be present, as had at least one enthusiast from Sweden.

As the appointed time drew near, there was no sign of the rain easing but a crowd of committed spectators gathered around the as-yet-unseen statue, sculpted by Whiteabbey-based artist John Sherlock, one of Northern Ireland's leading sculptors and an internationally recognised artist. His previous local commissions have included a life-size bronze statue of Professor Frank Pantridge, inventor of the cardiac defibrillator, and another great Ulsterman.

Right on cue, the dignitaries, led by Eric Jess, Chairman of the Committee and Sally Fleming, Harry's granddaughter, arrived at the memorial and, following brief speeches, the covers were off.

It was obvious, given the reaction of the crowd, that John Sherlock had captured the essence of the great man and spontaneous applause broke out. Eric Jess voiced the sentiments of all present when he finished his comments by saying: "Harry is home". ■

Approaching lunchtime and the great man is still under wraps.

Perhaps the rain was appropriate to the obviously contemplative mood captured by the sculptor.

3

4

5

READY TO PAINT

Country Landscapes in Watercolour

Terry Harrison

SEARCH PRESS

First published in Great Britain 2011

Search Press Limited
Wellwood, North Farm Road,
Tunbridge Wells, Kent TN2 3DR

ISBN: 978-1-84448-643-4

Suppliers
If you have any difficulty obtaining any of the materials and equipment mentioned in this book, please visit the Search Press website: www.searchpress.com or Terry Harrison's website: terryharrison.com.

Publisher's notes
All the step-by-step photographs in this book feature the author, Terry Harrison, demonstrating watercolour painting techniques. No models have been used.

Please note that when removing the perforated sheets of tracing paper from the book, score them first, then carefully pull out each sheet.

Share your Ready to Paint artworks with the world. Simply go to the 'Ready to Paint' page on Facebook and upload your images.

Printed in China

Dedication

When I looked through the selection of paintings for this book, I was reminded of my loving old pet dog Molly, a bearded collie. Together we have visited all of the places shown here. She always thought it was just a walk in the country, though to me, it was my job – what a way to make a living! So I dedicate this book to the memory of Molly.

Acknowledgements

It's not all about me all of the time, I do need help (no, not that type of help!) with the admin side of my books, so my thanks go to my editor, Sophie Kersey for doing another splendid job.
Also my thanks go to my wife and typing angel, Fiona Peart.

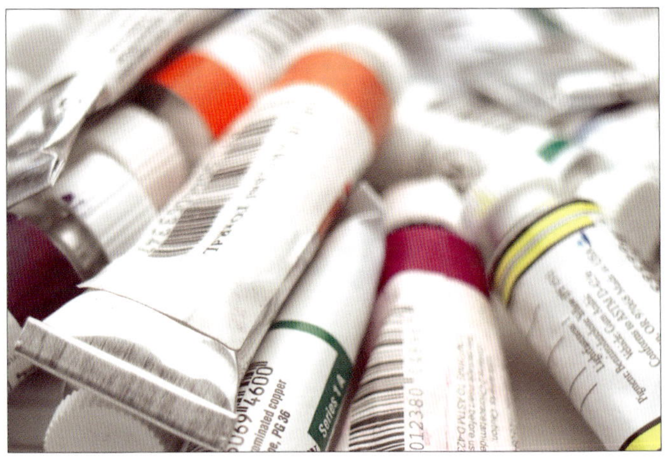

Page 1

Cherry Blossom Cottage

47 x 32cm (18½ x 12⁵/₈in)

Using many of the methods demonstrated in the Blossom Meadow project on pages 26–33, I have created this April landscape featuring a thatched cottage. Masking fluid was used on the cherry trees to achieve the texture of the blossom, then permanent rose was stippled over the area when the masking fluid was removed. The two silver birches on the right of the painting were also masked with masking fluid, as were the yellow flowers along the grass verge. The road sweeps past the cottage then down into a dip, and reappears further on, leading you into the distance. This feature is called 'lost and found'.

Opposite

The Bluebell Cutting

52 x 31cm (20½ x 12¼in)

Deep in the Hampshire countryside winds the Basingstoke Canal (which doesn't actually reach Basingstoke), through mile upon mile of unspoilt landscape just waiting to be painted. I have spent many happy hours walking alongside and later painting this canal. I would like to share with you this stretch of canal where in the spring the banks are awash with bluebells.

Contents

Introduction

I often hear people say they would love to paint but they can't draw, and many amateur artists are let down by their drawing skills. There is also the problem of choosing the right subject to paint, and then there is the composition – it's enough to make you give up! Help is at hand in this book, as it removes the first hurdle, the drawing, so that you are free to follow the easy step-by-step instructions to a successful painting. With success comes confidence, which leads to bigger and better things. In this book are five project paintings with a country landscape theme, and for each painting there is a line drawing printed on tracing paper. There is also a bonus tracing of the painting opposite. Follow the instructions on page 9 on how to transfer the images on to watercolour paper, and you are ready to paint.

I love painting country landscapes. For this book I have chosen some of my favourite views that I have stumbled across on trips out in the country (I always take a camera with me in case this happens). In fact, all of the paintings are inspired by views from a footpath or track, as if you have just discovered them on your journey.

To get the most out of this book, you can use the tracings over and over again. Each project can be repainted in different seasons, such as the project Snow Tracks on pages 34–39, which is transformed into a summer landscape on page 48. Use the same tracing, and for help with the summer colours use the Open Gate project on pages 10–17 as a guide. You can also swap elements in projects, for instance the church in the Short Cut project (pages 18–25) can be replaced with the barn in the Open Gate project. In this way you can create your own versions of the paintings. Remember, using tracings is not cheating, it is a traditional technique used by many professional artists, it's just their little secret!

Over the Hedgerow

45 x 33cm (17¾ x 13in)

This distant village scene with rustic buildings clustered around the church is framed by the overhanging tree and the shadow it casts over the road. Bathed in bright sunlight are a host of summer flowers that add colour and warmth to this fabulous country view.

TRACING 6

Materials

Paper

You can buy watercolour paper in three surfaces: Hot Pressed or smooth; Not, which has a bit of texture, and Rough which, as its name suggests, is more highly textured. I have used Rough paper for all the demonstrations in this book, as it is useful for some of the textured effects I need for landscapes.

Watercolour paper also comes in different weights, and the lighter weights tend to cockle when you apply watercolour washes. One way to prevent this is to stretch paper by soaking it and taping it to your painting board, so that when it dries, it is stretched flat. I prefer to use a heavier paper that is much less likely to cockle when wet. All the demonstrations in this book were done on 300gsm (140lb) watercolour paper.

Paints

There are two grades of watercolour paint: artists' quality and students' quality. Many beginners opt for the students' quality because they are cheaper. I would advise you to buy the best paints you can afford. Artists' quality watercolours have a higher pigment content, and they flow better and produce more vibrant results.

The other choice is whether to buy tubes or pans. When working in the studio, painting large washes and using bigger brushes, it is best to use tube colours. Pans can be more suitable if you are painting outdoors, as they come in compact sets that are more convenient outside. I have used artists' quality tube colours for all the demonstrations in this book.

You do not need a huge selection of colours for landscapes. I use a limited palette and mix all the colours from this range. For the paintings in this book, I used French ultramarine, burnt umber, raw sienna, cadmium yellow, burnt sienna, cadmium red, permanent rose, alizarin crimson and cobalt blue. I also used some of my own brand of colours: three greens and a shadow colour, but you can replace these with your favourite brands. They are: country olive (or olive green), midnight green (or Hooker's green), sunlit green (or green gold) and shadow (or French ultramarine, burnt sienna and permanent rose).

Brushes

I have used brushes from my own range in this book but you can use brushes from other ranges, as long as you look out for the qualities described. I used the golden leaf brush for large washes, as it holds a lot of paint, but you could use another large wash brush. I used my fan gogh, which is a thick fan-shaped brush; you could use another fan brush. The 19mm (¾in) flat brush was used for painting water and dragging down reflections. My large, medium and small detail brushes are good for general use: they hold varying amounts of paint and all come to a good point. No. 12, no. 8 and no. 4 round brushes could be used instead. My foliage brush is useful for stippling, but you could use a small bristle brush instead. I use a half-rigger for fine detail, which is like a rigger but shorter, and a special masking fluid brush to avoid spoiling my good brushes.

From left to right: my foliage brush, half-rigger, masking fluid brush, small, medium and large detail brushes, 19mm (¾in) flat, fan gogh and golden leaf brush.

Other materials

I attach my watercolour paper to a **painting board** with **masking tape**.

I lay out my tube colours in a plastic **palette** and mix washes in the wells.

Use spare watercolour paper or pieces cut from a magazine to make a **paper mask**.

Masking fluid is used to keep areas white. It is applied with a brush or with a **ruling pen** for straighter lines. Wet the brush and coat it with **soap** before using masking fluid, and the fluid will wash out easily after use.

A **pencil** and **eraser** are used when transferring the image, and you can also use the handle of a **spoon**.

A **hairdryer** can be used to speed up the drying process.

Kitchen paper is handy for mopping up spills and for lifting out paint while it is wet.

Transferring the image

Tracings are provided at the front of this book for all the demonstration paintings. Follow the steps shown below to transfer the images on to watercolour paper. You can reuse each tracing several times so you can use the same basic image to produce two quite different paintings.

1 Turn the tracing over and Draw over the back of the lines with a 2B pencil. I have used a mechanical pencil.

2 Tape the tracing face-up on your watercolour paper with masking tape. Draw carefully over the lines with the pencil, to transfer the graphite on to the paper. You can also use the handle of a spoon.

3 Lift up the tracing to reveal the image transferred on to the watercolour paper ready for you to begin painting.

The Open Gate

TRACING 1

If you are looking for a subject to fill the foreground and lead the viewer into a painting, an open gate is ideal. A stile is another way of inviting the viewer to venture further into a painting. This summer landscape gives you an opportunity to use the paper mask technique for distant trees and hedgerows. The middle distant trees frame the farmhouse and are painted using the golden leaf brush. The woodwork of the fence and the gate in the foreground is masked with masking fluid, and the greens behind the fence and gate are painted very dark, so that when the masking fluid is removed, a strong contrast is created. The lighter detail stands out from the dark background.

Having a road in the centre of the painting leads you towards the focal point, but it could divide the composition into two halves. To overcome this, place some shadows across the road at the bottom of the painting. This will link one side with the other and help to frame the painting.

You will need

300gsm (140lb) Rough watercolour paper

Colours: French ultramarine, burnt umber, raw sienna, burnt sienna, shadow (French ultramarine, burnt sienna and permanent rose), midnight green (Hooker's green), cobalt blue, country olive (olive green), sunlit green (green gold), cadmium red

Brushes: golden leaf (large wash brush), large detail (no. 12 round), foliage (bristle brush), medium detail (no. 8 round), half-rigger (rigger), small detail (no. 4 round)

Masking fluid, masking fluid brush and ruling pen

Paper mask

1 Transfer the scene on to watercolour paper. Apply masking fluid to the fence, flower heads and gate with a brush, then use a ruling pen to create grasses and stalks in the foreground.

2 Use the golden leaf brush to wet the sky area with clean water, then paint on French ultramarine, leaving spaces for clouds. Mix ultramarine and burnt umber and paint cloud shadows while the painting is wet.

3 Use the large detail brush to wet the track area with clean water, then drop in very pale raw sienna from the far end, strengthening the mix as you come forwards. Drop in burnt sienna and raw sienna in the foreground.

4 Still working wet in wet, indicate the furrows in the track with shadow colour. Allow to dry.

5 Make a pale mix of midnight green and cobalt blue and stipple along the edge of a paper mask with the foliage brush to suggest trees on the horizon.

6 Paint the right-hand part of the horizon in the same way, then move the paper mask at an angle so that you can stipple hedgerows bordering fields coming forwards.

7 Keep changing the angle of the paper mask and as you come forwards, use a stronger mix of country olive and cobalt blue.

8 Stipple the hedge in front of the house with a light mix of country olive and cobalt blue, then while this is wet, stipple a darker colour at the base of the hedge with midnight green. Continue this hedge in the same way to the left and right.

9 Use the medium detail brush to drop individual colours into the fields, starting with pale midnight green in the distance, then sunlit green, then raw sienna.

10 Use the golden leaf brush with a mix of raw sienna and sunlit green to paint the field in the middle distance. Add burnt sienna on the right.

11 Mix country olive and sunlit green and paint this over the masked gate and fence area.

12 Paint midnight green further forwards, leaving gaps on both sides, then fill the gaps with sunlit green, suggesting a patch of sunlight.

13 Stipple along the edge of the track into the distance with midnight green. Change to the foliage brush and stipple country olive along the right-hand edge of the track.

14 Use the half-rigger and midnight green to paint the trunks and branches of the trees.

15 Stipple the trees using the foliage brush, starting with sunlit green and then adding midnight green for the shaded right-hand sides.

16 Place a paper mask over the edge of the barn roof and stipple behind it, then do the same for the roof of the house. Use the small detail brush to tidy the edges.

17 Mix a blue-green from cobalt blue and sunlit green and paint a tree in front of the house. Add midnight green on the right-hand side.

18 Stipple the foliage of the trees on the right with raw sienna and sunlit green on their lit left-hand sides, then with midnight green and burnt sienna on the right.

19 Make a pale mix of burnt sienna and paint the roofs with the medium detail brush.

20 While this is still wet, drop in the shadow colour, which will darken the roofs and add texture.

21 Use the small detail brush and a darker mix of burnt sienna to paint the shaded parts of the house.

22 Make a dark mix of ultramarine and burnt umber and paint the windows and other architectural details of the house.

23 Paint the lit side of the barn with a pale mix of burnt umber and the medium detail brush, then paint the front with the same mix and drop in a mix of burnt umber and country olive while the first wash is wet. Allow to dry.

24 Use the small detail brush and a mix of ultramarine and burnt umber to paint the shadow under the eaves and the weatherboarding.

25 Change to the golden leaf brush and stipple plants over the masked gate and fence with midnight green.

26 Paint grasses in the foreground with the half-rigger and midnight green.

27 Use the large detail brush and shadow colour to paint shadows across the track from trees that are outside the scene. Allow to dry.

28 Remove the masking fluid by rubbing with your fingers. Make a pale mix of burnt sienna and ultramarine and use the large detail brush to paint the fence and gate. Allow to dry. Use the medium detail brush to paint shade with a darker mix.

29 Make a mix of ultramarine and burnt umber and paint dappled shadow over the gate.

30 Wash over the grasses that were masked with sunlit green.

31 Shade the undersides of the cow parsley flowers with a thin wash of cobalt blue.

32 Paint the flowers that were masked with cadmium red to create poppies, then paint more with the tip of the brush in the distance.

33 Add darker grasses in the foreground with the half-rigger and midnight green.

34 Paint a glaze of sunlit green over the dried paint of the blue-green tree in front of the house, and over white gaps in the other trees.

Overleaf

The finished painting.

The Short Cut

TRACING 2

This is a very simple but effective composition. The church is the focal point and this is framed by the large trees in the foreground on the left, with the bluebells and shadows running horizontally along the bottom of the painting and the tree on the right completing the frame. The stile and signpost guide the viewer into the painting, and the footpath leads you through the stile and beyond. If you are in any doubt which way to go, just follow the signpost!

Masking fluid on the tree trunks helps to retain the light and shade and gives the impression of a bright, sunny spring day. If you feel like a change, you can easily substitute the church for a barn or a farmhouse from one of the other projects.

You will need

300gsm (140lb) Rough watercolour paper

Colours: French ultramarine, burnt umber, raw sienna, sunlit green (green gold), midnight green (Hooker's green), cobalt blue, country olive (olive green), permanent rose

Brushes: golden leaf (large wash brush), foliage (bristle brush), medium detail (no. 8 round), half-rigger (rigger), large detail (no. 12 round), small detail (no. 4 round)

Masking fluid, masking fluid brush and ruling pen

Paper mask

1 Use masking fluid and a brush to mask the trees, stile and signpost, then mask the grasses with a ruling pen.

2 Wet the sky area with the golden leaf brush and clean water, then paint the blue with ultramarine, leaving space for clouds. Mix ultramarine and burnt umber to paint darker areas, then allow to dry.

3 Paint a wash of raw sienna behind the trees on the left, then add sunlit green wet in wet, before the first wash dries.

4 While this is wet, drop in midnight green to suggest darker foliage.

5 Use a paper mask and the folliage brush to stipple the trees on the horizon with cobalt blue and midnight green. Change the angle of the paper mask and stipple a stronger, greener mix to suggest hedgerows coming forwards.

6 Mix country olive and cobalt blue for a warmer green, and use the paper mask horizontally to paint the middle distant hedgerow. While this is wet, stipple midnight green at the bottom.

7 Use the medium detail brush to paint the distant fields with a pale mix of midnight green and cobalt blue, then raw sienna coming further forwards.

8 Paint the right-hand distant field with sunlit green, then mix sunlit green and raw sienna to paint the middle distant field.

9 While the first wash is wet, paint shadows from the trees in the field with midnight green.

10 Paint the distant tree trunks with the half-rigger and a mix of burnt umber and country olive. Allow to dry, then paint on a wash of sunlit green for foliage with the foliage brush.

11 Stipple lower down with midnight green. Mask the edge of the church with a paper mask to preserve the edge.

12 Use the paper mask to create a straight line along the bottom of this wooded area and stipple midnight green, then country olive to suggest undergrowth here.

13 Paint negatively around the church to re-establish the edges with the medium detail brush and country olive.

14 Stipple texture on to the trees with midnight green.

15 Stipple sunlit green on the three trees on the right, then stipple on midnight green, using a paper mask to protect the edge of the church.

16 Paint the footpath with raw sienna and the large detail brush, then drop in burnt umber wet in wet for texture.

17 Use the medium detail brush and a mix of midnight green and cobalt blue to paint the trees behind the main trees on the left.

18 Paint the main tree trunks over the masking fluid with the large detail brush and burnt umber and country olive, then add ultramarine and paint the darker mix down the right-hand sides.

19 Using the golden leaf brush and midnight green, stipple downhill from the trees to the path.

20 Stipple patches of sunlit green to suggest sunlight filtering through the trees.

21 Mix cobalt blue and permanent rose and stipple the bluebells on to the woodland floor on both sides of the path. Allow to dry.

22 Use the foliage brush to stipple on a stronger mix.

23 Stipple midnight green among the bluebells, suggesting grasses and undergrowth.

24 Stipple and flick up grasses with midnight green along the bottom of the fence.

25 Darken the area around the church with the medium detail brush and midnight green. Change to the foliage brush and stipple a little texture with a pale mix of burnt umber and a touch of cobalt blue.

26 Add more cobalt blue to the mix and use the medium detail brush to add shadows and define edges, giving the building shape. Allow to dry.

27 Use the small detail brush to paint the roofs with burnt sienna, then drop in cobalt blue and burnt sienna while this is wet.

28 Paint the shed beside the church with a pale mix of burnt umber, then paint the roof and right-hand side with a strong mix. Use this to paint the architectural details of the church.

29 Remove the masking fluid with clean fingers and paint the left-hand trees, the lit parts of the larger trees and the fence posts, signpost and stile with a pale mix of country olive and burnt umber using the large detail brush.

30 Use the medium detail brush and a darker mix of country olive and burnt umber to add shade to the right-hand sides of the trees, and to add more branches. Add shadows across the main tree trunks in the same way.

31 Add shade to the fence posts, signpost and stile with the same mix.

32 Paint a wash of sunlit green over the grasses that were masked, on both sides of the path.

33 Use midnight green to paint dark shadows coming from the trees on the left, and to darken the bottom left-hand corner.

34 Re-establish texture in the area of the bluebells with the foliage brush and midnight green.

35 Use the medium detail brush and shadow colour to paint the tree shadows going over the path.

36 Stipple more texture in the trees with the golden leaf brush and country olive, then midnight green.

37 Paint the fence wire with the half-rigger and burnt umber.

Overleaf
The finished painting.

Blossom Meadow

TRACING 3

This riverside footpath reminds me of many treks along riverbanks, looking for a subject to paint. Here, I was suddenly treated to a scene that has everything I look for in a painting. The tree in blossom is the focal point, and the lazy river leads you into the painting and the distant meadows with a glimpse of the faraway hills beyond.

This project gives you the opportunity to create distance in a painting with plenty of light bluey tones for the faraway trees and hedgerows, and the pale, soft shades of the hills on the horizon. Masking fluid is used on the blossom and on the foreground flowers and grasses.

You will need

300gsm (140lb) Rough watercolour paper

Colours: French ultramarine, raw sienna, burnt sienna, cobalt blue, midnight green (Hooker's green), sunlit green (green gold), country olive (olive green), burnt umber, cadmium yellow, permanent rose

Brushes: golden leaf (large wash brush), large detail (no. 12 round), medium detail (no. 8 round), foliage (bristle brush), small detail (no. 4 round), half-rigger (rigger), 19mm (¾in) flat

Masking fluid, masking fluid brush and ruling pen

Paper mask

1 Mask the blossom and flower heads on the ground with a brush, then use the ruling pen to apply masking fluid to the grasses.

2 Wet the sky area with the golden leaf brush and clean water, then paint it with ultramarine, leaving white paper for clouds. While this is wet, drop a thin wash of raw sienna into the clouds.

3 Paint the footpath with the large detail brush and raw sienna. Allow to dry.

4 Paint the distant hill on the far right with the medium detail brush and pale ultramarine, then to the left of this paint ultramarine and drop in burnt sienna. Further forwards, paint a pale mix of cobalt blue and midnight green.

5 Use a paper mask to create a straight hedgerow and stipple a stronger mix of midnight green and cobalt blue above this with the foliage brush. Repeat on the left of the painting, creating trees and bushes in the hedgerow.

6 Use the medium detail brush to drop various colours into the distant fields: sunlit green in some and watered down country olive in others. Change to the large detail brush and paint raw sienna coming forwards.

7 While this field is wet, drop in country olive. Allow to dry.

8 Mask the bottom of the hedgerow with the paper mask and use the foliage brush to stipple the lower part of it with midnight green.

9 Use the small detail brush to paint tree trunks and branches with midnight green. Allow to dry, then paint the distant hills behind the trees with a pale wash of cobalt blue.

10 Using the foliage brush, stipple a mix of sunlit green and cobalt blue on to the trees, then midnight green.

11 Paint the trunks and branches of the trees to the right of these with the half-rigger and burnt umber and midnight green, then stipple on foliage with the golden leaf brush and ultramarine with country olive. Mix midnight green and ultramarine to stipple on darker foliage.

12 Stipple the lighter tree to the right with sunlit green, then continue to the right of this with midnight green and country olive. Allow to dry.

13 Use the foliage brush to paint the field on the left of the painting with raw sienna, then country olive further forwards, then stipple midnight green wet in wet down the river bank.

14 Stipple sunlit green foliage at the top of the tree in blossom, then stipple with midnight green on the shaded side, over the masking fluid.

15 Stipple pale country olive on the left-hand edge of the river bank, then stipple and flick up midnight green on the far left, wet in wet. Allow to dry.

16 Using the medium detail brush, paint the main tree trunk and branches, going up into the foliage, with country olive and burnt umber. Shade the right-hand side with a darker mix and add more branches.

17 Stipple more foliage with the golden leaf brush and midnight green.

18 Paint grass under the main tree with the foliage brush and pale country olive, then add paler grasses with raw sienna.

19 Paint a mix of country olive and midnight green along the base of the fence, suggesting grasses.

20 Make a pale mix of raw sienna and country olive and paint this on either side of the path, then drop in a stronger mix of country olive.

21 Paint on either side of the path, over the masked flower heads, with country olive and burnt umber, then paint with sunlit green and raw sienna further forwards.

22 Flick up grasses in the foreground with midnight green, then stipple very pale burnt umber on to the path.

23 Flick up and stipple midnight green down the river bank on the left, to add texture.

24 Stipple more texture on either side of the path in the same way.

25 Use the 19mm (¾in) flat brush to drag a pale wash of cobalt blue downwards to paint the water.

26 Add country olive to the mix and drag this down to create reflections, then continue with cobalt blue on the left.

27 Drag down a little cobalt blue and country olive on the right-hand side, then use midnight green to reflect the darker area at the water's edge.

28 Use the small detail brush and midnight green to paint the dark reflections at the edge of the right-hand bank.

29 Paint reflections of the main tree and the more distant tree with country olive and burnt umber. Allow to dry.

30 Rub off the masking fluid and paint the grasses with a wash of sunlit green on the medium detail brush.

31 Paint the flower heads with cadmium yellow.

32 Make a pale mix of burnt umber and ultramarine and paint the fence posts and stile. Allow them to dry, then shade the right-hand sides with a darker mix.

33 Paint the blossom with permanent rose, using a stronger mix in places.

34 Paint the reflection of the blossom with permanent rose, and drag the colour down.

35 Use midnight green to paint the darker part of the reflection of the blossom bush and the edge of the bank.

36 Paint the shadow from the main tree with the small detail brush and midnight green.

37 Use the half-rigger to paint the fence wires with burnt umber.

38 Finally use the foliage brush and midnight green to stipple more texture on the trees.

Overleaf

The finished painting.

Snow Tracks

TRACING 4

Masking fluid is a perfect medium to use when painting snow scenes, but care must be taken when applying it, as it dries very quickly and can ruin your brushes. Use a synthetic brush for masking fluid and see page 8 for information on how to use soap to protect it.

The dusting of snow on the trees is achieved by making a pad of kitchen paper, dipping it into some masking fluid, then very lightly tapping it over the trees. Practise this before starting the painting, and if it is not quite right on your first attempt, wait for it to dry, rub it off, then start again. If you fancy painting this scene as a summer landscape, as seen on page 48, you can use the colours and techniques used in the Open Gate project on pages 10–17.

You will need

300gsm (140lb) Rough watercolour paper

Colours: French ultramarine, burnt umber, burnt sienna, shadow (ultramarine, burnt sienna and permanent rose), cobalt blue

Brushes: golden leaf (large wash brush), large detail (no. 12 round), medium detail (no. 8 round, foliage brush (bristle brush), half-rigger, fan gogh (fan brush)

Masking fluid, masking fluid brush and ruling pen

Kitchen paper

Paper mask

1 Apply masking fluid to the fence, the tracks in the snow and some of the branches with a brush, then use a ruling pen to apply it to other, finer branches. Pick up some masking fluid on scrunched up kitchen paper and apply this for snow in the tree tops.

2 Wet the sky area with the golden leaf brush and clean water, being careful not to go over the roof lines, then drop in ultramarine. While this is wet, add clouds with ultramarine and burnt umber. Allow to dry.

3 Paint the top of the gable end of the barn with the large detail brush and a strong, dark mix of ultramarine and burnt umber, then while this is wet, paint a pale, browner mix lower down.

4 Paint the barn door with the same mix, then drop in the darker version at the top. Make a pale mix of ultramarine and burnt umber and use the medium detail brush to paint the side of the porch.

5 Paint the redder parts of the barn with burnt sienna, then drop in shadow colour wet in wet.

6 Use a paper mask to mask the horizon and use the foliage brush to stipple distant trees and bushes with cobalt blue and shadow colour.

7 Place the paper mask at various angles as you come forwards and paint the hedgerows with a stronger mix of the same colours, adding burnt umber for warmth.

8 Use the medium detail brush and a dark mix of burnt umber and ultramarine to paint the trunks and branches of the trees, going over the masking fluid in places.

9 Make a fairly blue mix of ultramarine and burnt umber and lightly stipple winter foliage over the masking fluid using the foliage brush.

10 Paint the trunk of the large tree on the left with the large detail brush and a pale mix of burnt umber and ultramarine, then while this is wet, drop in a darker mix on the right to create a three-dimensional look.

11 Use the medium detail brush and the dark mix to paint the branches under the masked out snow.

12 Paint a wash of cobalt blue over the shadowed parts of the snowy roofs with the large detail brush. Make the gable end triangles a slightly stronger blue.

13 Use a half-rigger and a mix of ultramarine and burnt umber to paint planking, adding dark shapes to suggest broken planks.

14 Give form to the tree to the right of the barn by adding shadow with the medium detail brush and ultramarine and burnt umber.

15 Paint a wash of cobalt blue on the snow in front of the barn, over the masking fluid, indicating the shape of the snow and the land.

16 Mix pale cobalt blue with a touch of burnt sienna and paint the shadows in the tracks in the snow, then with cobalt blue, paint shadows from the trees across the path, and in the undergrowth on the right.

17 Paint the fence posts with ultramarine and burnt umber, starting with the uprights and then painting under the masking fluid.

18 Use the fan gogh brush with burnt umber to flick up grasses showing through the snow.

19 Add grasses and texture on the right-hand side of the painting, holding the fan gogh brush vertically to suggest dried seed heads.

20 Use the medium detail brush and cobalt blue to paint the shadows of the fence posts.

21 Add texture in the undergrowth on the right with a darker wash of cobalt blue.

22 Paint more branches and twigs in the tree on the left with the half-rigger and ultramarine and burnt umber, then paint twigs in the undergrowth on the right in the same way.

23 Change to the medium detail brush and paint a pale wash of cobalt blue in the distant fields.

24 Mix cobalt blue with burnt sienna and paint stronger shadows in the tracks in the snow. Allow to dry.

25 Remove the masking fluid. Use the large detail brush to paint a wash of cobalt blue over the snow in the trees, leaving white at the top.

26 Shade under the snow on the fence in the same way.

27 Continue the shadow of the tree across the foreground.

Overleaf

The finished painting.

New Forest Bridge

TRACING
5

Autumn in the New Forest is a magical time of year; the colours are amazing. This footbridge was in a part of the forest called Pig's Bush, which is an unfortunate name for such a beautiful location. The bridge has recently been replaced with a newer version, but the original is a much more appealing subject to paint.

Most of the tree trunks are quite dark and are silhouetted against the golds and light reds of the foliage background. The shadow across the track at the bottom of the painting helps to frame the focal point of the bridge. This scene can easily be transformed into a summer landscape by using greens for the foliage: sunlit green replaces the gold colours, country olive replaces the reds, and midnight green is a substitute for the dark browns.

You will need

300gsm (140lb) Rough watercolour paper

Colours: French ultramarine, cadmium yellow, raw sienna, cadmium red, cobalt blue, alizarin crimson, burnt sienna, sunlit green (green gold), burnt umber, country olive (olive green), shadow (French ultramarine, burnt sienna and permanent rose)

Brushes: golden leaf (large wash brush), foliage brush (bristle brush), large detail (no. 12 round), medium detail (no. 8 round), half-rigger (rigger), fan gogh (fan brush)

Masking fluid, masking fluid brush and ruling pen

Paper mask

1 Apply masking fluid to the bridge, the ripples in the water and the lit sides of some tree trunks with the brush, then use the ruling pen to mask the grasses.

2 Use the golden leaf brush and clean water to wet a small area of sky, and drop in ultramarine to create a patch of blue.

3 Stipple a bright area of foliage on the right-hand side with cadmium yellow and raw sienna, then while this is wet, stipple cadmium red and cadmium yellow on top, then on the left of the painting. Allow to dry.

4 Use a paper mask and the foliage brush with a mix of cobalt blue and a touch of alizarin crimson to stipple distant trees. While this is wet, stipple cadmium red and cadmium yellow on to the sunlit left-hand side.

5 In the same way, stipple the lower parts of the left-hand woods with cobalt blue, alizarin crimson and burnt sienna, then mask the bridge rail and stipple the same mix on the right.

6 Stipple into the distant trees with a mix of cadmium yellow with a touch of cadmium red, then repeat with burnt sienna.

7 Stipple across the foliage using the golden leaf brush with cadmium yellow, then with more burnt sienna.

8 Still working wet in wet, stipple sunlit green lower down on the right.

9 Mix a dark purple-brown from burnt umber, alizarin crimson and ultramarine. Mask the body of the bridge with paper and stipple behind it.

10 Use the large detail brush to paint the track and the path over the bridge with a wash of raw sienna. Allow to dry.

11 Paint the middle ground with the medium detail brush and burnt sienna, and while it is wet, drop in burnt umber.

12 Use the golden leaf brush to stipple and flick up undergrowth in the left-hand corner with burnt sienna, then while this is wet, paint a dark mix of ultramarine, alizarin crimson and burnt umber at the very bottom.

13 Stipple more of the same dark mix around the painting to add texture, particularly round the edges to frame it.

14 Mix ultramarine and burnt umber and use the medium detail brush to paint trunks and branches. Refer to the tracing if you have obscured some of the pencil lines with paint.

15 Use the half-rigger to paint the trunks and branches of the distant trees.

16 Add country olive to the mix and paint the foreground trees with the medium detail brush, painting over the masking fluid. Allow to dry.

17 Rub off the masking fluid, apart from the ripples in the water. Mix raw sienna and sunlit green and paint over the masked out, highlighted parts of the trees. While this is wet, drop in some of the previous colour, which will blend and soften the edges.

18 Use the large detail brush with a pale mix of sunlit green and burnt umber to paint the woodwork of the bridge. Paint the inside part in deep shade with a dark mix of ultramarine, burnt umber and country olive. Allow to dry.

19 Paint the shaded detail and the planking of the bridge with the same mix and the medium detail brush.

20 Make a very dark mix of ultramarine and burnt umber and paint the shadow under the bridge.

21 Stipple texture on the track with the foliage brush and burnt umber.

22 Use the fan gogh brush and a mix of burnt umber and country olive to paint grasses at the base of the bridge.

23 Change to the medium detail brush and shadow colour and paint shadows across the track and up the bridge.

24 Use the large detail brush to paint a shadow across the foreground, supposedly from a tree outside the picture. Make it dappled to imply foliage.

25 Paint the water with the large detail brush and ultramarine, then while this is wet, paint reflections with raw sienna and burnt sienna.

26 Mix ultramarine, burnt umber and country olive and paint the reflection of the bridge's dark underside, and ripples. Allow to dry.

27 Dilute the mix and use the medium detail brush to paint lighter reflections and ripples. Then paint the back of the bridge showing through the arch with burnt umber and ultramarine, and add a reflection and ripples.

28 Add a pale hint of raw sienna and burnt sienna to the rippled reflections.

29 Paint some of the remaining white areas in the foliage with a wash of cadmium yellow on the medium detail brush.

30 Paint a very pale wash of burnt sienna over the masked grasses.

31 Darken the handrail and uprights at the back of the bridge with very pale burnt umber.

32 The reflection of the foreground tree should just be visible behind the grasses, so paint this in with ultramarine and burnt umber, then paint the reflection of the bridge upright. Remove the masking fluid from the water.

33 Paint smaller branches and twigs across the sky area with the half-rigger and a mix of burnt umber and ultramarine.

Overleaf

The finished painting.

Index

Summer Tracks

45 x 34cm (17¾ x 13⅜in)

Using the same tracing as the Snow Tracks project on pages 34–39, I have created a summer version of the painting. All the techniques you need are demonstrated in the Open Gate project on pages 10–17.